ADVANCES IN OPTOELECTRONIC TECHNOLOGY AND INDUSTRY DEVELOPMENT

T0133450

Symposium on Photonics and Optoelectronics: proceedings

The International Symposia on Photonics and Optoelectronics (SOPO) aim to provide a premier technical forum for researchers, engineers as well as professionals from all over the world to present the latest research and development in Photonics and Optoelectronics related fields. Optoelectronic systems use laser technology, for a wide range of applications in the fields of telecommunications, electronic industrial, civil, and aeronautical engineering. SOPO features the latest advances in optoelectronic integration, laser technology, manufacturing of optical components, advanced laser materials, films and devices, optical communications, silicon photonics, quantum optics, optoelectronic devices and integration, medical and biological applications and image processing.

Print ISSN: 2156-8464
Online ISSN: 2156-8480

PROCEEDINGS OF THE 12th INTERNATIONAL SYMPOSIUM ON PHOTONICS AND OPTOELECTRONICS (SOPO 2019), AUGUST 17-19, 2019, XI'AN, CHINA

Advances in Optoelectronic Technology and Industry Development

Edited by

Gin Jose
University of Leeds, UK

Mário Ferreira
University of Aveiro, Aveiro, Portugal

CRC Press
Taylor & Francis Group
Boca Raton London New York

CRC Press is an imprint of the
Taylor & Francis Group, an **informa** business

A BALKEMA BOOK

Published by:
CRC Press/Balkema
P.O. Box 447, 2300 AK Leiden, The Netherlands
e-mail: Pub.NL@taylorandfrancis.com
www.crcpress.com – www.taylorandfrancis.com

First issued in paperback 2022

Publisher's Note
The publisher has gone to great lengths to ensure the quality of this reprint but points out that some imperfections in the original copies may be apparent.

Visit the Taylor & Francis Web site at
http://www.taylorandfrancis.com

and the CRC Press Web site at
http://www.crcpress.com

Typeset by Integra Software Services Pvt. Ltd., Pondicherry, India

Although all care is taken to ensure integrity and the quality of this publication and the information herein, no responsibility is assumed by the publishers nor the author for any damage to the property or persons as a result of operation or use of this publication and/or the information contained herein.

ISBN 13: 978-1-03-240114-0 (pbk)
ISBN 13: 978-0-367-24634-1 (hbk)
ISBN 13: 978-0-429-28362-8 (ebk)

DOI: https://doi.org/10.1201/9780429283628

Advances in Optoelectronic Technology and Industry Development – Jose & Ferreira (eds)
© 2020 Taylor & Francis Group, London, ISBN 978-0-367-24634-1

Table of contents

Optical Communications

Optoelectronic Devices and Integration

Photonics and Optoelectronics

Preface

The fields of Photonics and Optoelectronics have grown spectacularly over the last few decades. Actually, these fields are intertwined and have a profound effect on the emergence of modern technologies and their influence on our lives. Experimental and theoretical research into such technologies include sources and detectors as components, active elements such as modulators/switches, integration of optical components to perform various functions, and the development of active optical system/subsystem technologies.

The International Symposium on Photonics and Optoelectronics (SOPO) has successfully brought the fields of Photonics and Optoelectronics together and has become an established meeting in China since 2009, involving a truly diverse set of well known experts from Asian Pacific areas, North America, Europe and around the world. It was held in Wuhan, Chengdu, Shanghai, Suzhou, Xi'an, Beijing, Guilin, Sanya and Kunming from 2009 to 2018. SOPO 2019 took place in Xi'an, China from August 17 to 19 and it was co-organized by Wuhan University, Beijing University of Posts and Telecommunications, IPOC State Key Lab, Liaocheng University and other institutions.

SOPO 2019 featured the latest advances in optoelectronic integration, laser technology, manufacturing of optical components, advanced laser materials, films and devices, optical communications, silicon photonics, quantum optics, optoelectronic devices and integration, medical and biological applications and image processing. The conference included 22 Plenary Speeches and many other regular communications. However, due to Ei indexing requirements, only a limited number of papers have been selected for publication. Actually, the 36 papers included in this Proceedings provide an excellent overview of the topics presented at SOPO 2019. Among these papers, 8 are dedicated to optical sensors, 7 are in the area of optical communications, 7 are related with imaging and detection, 2 are in the area of quantum optics, 6 are concerned with lasers and other optical sources, while the remaining 6 papers are dedicated to different types of optoelectronic and photonic devices. We are sure that this publication will arouse great interest both from specialists in these fields and from general readers.

SOPO 2019 could not have happened without the hard work of many people, namely those integrating the Local Organizing Committee and the Session Chairs, who helped in many ways to assemble and run the conference, the Technical Program Committee and, especially, the General Chair, Prof. Zhiping Zhou, who guided the technical direction of the conference and assisted with the conference program and proceedings. We are also extremely pleased to have received generous sponsorship from several organizations. Finally, we are also very grateful to the Editorial Staff at CRC Press/Balkema, and the many contributors and reviewers for helping us to put-together this Proceedings

Mário F. S. Ferreira
Main Editor
Department of Physics, University of Aveiro, Aveiro, Portugal

Organizers

Wuhan University
Beijing University of Posts and Telecommunications
IPOC State Key Lab
Liaocheng University

GENERAL CHAIR

- Prof. Zhiping Zhou
Peking University, China

TECHNICAL PROGRAM COMMITTEE

- Prof. Chong Leong Gan
Western Digital, Malaysia
- Prof. Yufei Ma
Harbin Institute of Technology, China
- Prof. Jietai Jing
East China Normal University, China
- Prof. Khaled Habib
Kuwait Institute for Scientific Research (KISR), Kuwait
- Prof. Yang Yue
Nankai University, China
- Prof. Shengjun Zhou
Wuhan University, China
- Prof. Igor V. Minin
FGUP "SNIIM", Novosibirsk/Tomsk Polytechnical University, Tomsk, Russia
- Prof. Peyman Goli
Khavaran Institute of Higher Education, Mashhad, Iran
- Prof. Mário F.S. Ferreira
University of Aveiro, Portugal
- Prof. Xuewen Shu
Wuhan National Laboratory for Optoelectronics, China
- Prof. Xingjun Wang
Peking University, China
- Prof. Hassan Pakarzadeh
Shiraz University of Technology, Iran
- Prof. Wenjie Wan
Shanghai Jiao Tong University, China
- Prof. Yuheng Wang
Guangdong Provincial Engineering Research Centre on Solid-State Lighting and its Informationisation, South China University, China
- Prof. Jingsong Li
Anhui University, China
- Prof. Zabih Ghassemlooy
Northumbria University, UK
- Prof. Mohammed M. Shabat

Islamic University of Gaza, Palestine
• Prof. Zheng Shang Da
Xi'an Institute of Optics & Precision Mechanics of Chinese Academy of Sciences, China
• Dr. Kang Wei
Apple, Inc, USA
• Prof. Gin Jose
University of Leeds, UK
• Dr. Te Hu
Apple Camera Team, USA
• Dr. Lin Xu
University of Southampton, UK
• Prof. Jean-Luc ADAM
UMR CNRS 6226 Institut des Sciences Chimiques de Rennes, France
• Dr. Xiaotian Li
Grating Technology Laboratory, Changchun Grating Technology Laboratory, Chinese Academy of Sciences, China
• Dr. Gholamreza Shayeganrad
University of Southampton, UK
• Prof. Khatereh Khorsandi
Tehran University of medical sciences (TUMS) branch, Iran

Image processing

Advances in Optoelectronic Technology and Industry Development – Jose & Ferreira (eds)
© *2020 Taylor & Francis Group, London, ISBN 978-0-367-24634-1*

Electrically-modulated optoelectronics-based infrared source enabling ground surface precision deflectometry

Henry Quach
James C. Wyant College of Optical Sciences, University of Arizona, Tucson, AZ, USA

Logan R. Graves
James C. Wyant College of Optical Sciences, University of Arizona, Tucson, AZ, USA
Intuitive Optical Design Lab, Tucson, AZ, USA

Hyukmo Kang
James C. Wyant College of Optical Sciences, University of Arizona, Tucson, AZ, USA

Dae Wook Kim
James C. Wyant College of Optical Sciences, University of Arizona, Tucson, AZ, USA
Department of Astronomy, University of Arizona, Tucson, AZ, USA

ABSTRACT: We introduce the design of a scalable, modulated long-wave infrared source. The design makes use of a pseudo-blackbody heating element array, which radiates into a custom aluminum integrating cavity. The elements possess low thermal capacitance, enabling temporal modulation for improved signal isolation and dynamic background removal. To characterize performance, deflectometry measurements were made using both the new source design and a traditional tungsten ribbon source, which possess similar source irradiance and identical emission profile dimensions. Measurements from a ground glass flat and an aluminum blank demonstrated the new source produces a signal-to-noise ratio four times greater than that of the ribbon. Thermal imaging demonstrated improved source geometry and signal stability over time, and further, the new design measured a previously untestable hot aluminum flat (150 °C). The new design enables high-contrast thermal measurement of surfaces typically challenging to infrared deflectometry due to high surface roughness or intrinsic thermal noise generation.

Keywords: optical metrology, deflectometry, infrared source, temporal modulation

1 INTRODUCTION

Freeform optics, or non-rotationally symmetric, highly custom-shaped optics, present the opportunity to improve the performance of an optical system. Under the assumption that an arbitrary surface can be fabricated, freeform optical designs may benefit from improved mechanical compactness, reduced assembly complexity, and better imaging performance (Fang et al., 2013). Across astronomical optics, medical imaging, defense, and consumer electronics, a new generation of optical instruments are on the horizon thanks to the flexibility conferred by these optics. However, as the popularity and diverseness of freeform optical surfaces grow, so do the metrology requirements.

Non-contact metrology of freeform optics is commonly performed by interferometry and deflectometry (Kim et al., 2016). Interferometry is a null metrology method, requiring a null optic to use as a reference against the unit under test (UUT). Particularly for freeform surfaces, custom null optics are required. Diffractive elements called computer-generated holograms (CGH) are attractive null element options because they can null up to and including arbitrary surfaces with extreme or high-frequency features (Dubin et al., 2009). However, a CGH is typically designed for only one null configuration and may not always be a viable option for testing.

Alternatively, deflectometry is a non-null test method that has demonstrated comparable performance to interferometry over a range of freeform surfaces (Graves et al., 2019). In this method, an illumination source emits a known pattern, which specularly reflects off the UUT, and the reflected image of the source is captured by a camera. With precise knowledge of the setup's geometry, the local slopes of the UUT can be determined and integrated to generate a reconstructed surface map. Deflectometry carries the advantages of high testable slope dynamic range and measurement flexibility because it does not require a null reference. Of interest to optical fabrication, measurement of surface figure error during intermediary grinding phases may help prioritize local figure error correction so that a surface may more efficiently converge. Due to rapidly-changing surface profiles during grinding, CGHs are not practical, but deflectometry can become applicable if a camera and source of appropriate wavelength satisfy the specular reflectance condition (Oh et al., 2016; Lowman et al., 2018).

For optics in the grinding phase, which have a rough diffusing surface, a specular reflection can be achieved using long-wave infrared (LWIR) light, specifically, in the 7–14 μm region. LWIR cameras are readily obtainable, but LWIR thermal source options are currently limited. One common design is implemented by applying a current to a thin tungsten ribbon, which induces joule heating and creates a rectangular, pseudo-blackbody emitting source (Su et al., 2011, 2013; Oh et al., 2016). By scanning this rectangular ribbon in orthogonal directions, a line scanning source is created, and slope information can be obtained to construct a full aperture surface sag map. While this source enables infrared deflectometry testing of rough surfaces at a precision scale, several inherent characteristics including low signal-to-noise ratio (SNR), limited modulation depth, and low temporal stability, limit the testing accuracy and range.

We have created a new source design which addresses these prior issues and extends the applicable range of infrared deflectometry testing. The source is a Long-wave Infrared Time-Modulated Integrating cavity Source (LITMIS), which uses modular high efficiency and high stability resistive membrane pseudo-blackbody elements. A performance comparison was made against the traditional tungsten ribbon source, whose shape was identical to the exit slit of the box, using the same setup and LWIR camera. Results show promise in deflectometric testing of rough optics during their grinding phase and UUTs under thermal load.

2 TEMPORALLY MODULATED INFRARED SOURCE FOR DEFLECTOMETRY

2.1 *Deflectometry overview*

Deflectometry is a non-null technique that measures surface slopes of an optical surface, which are post-processed to reconstruct the surface. Dynamic range limitations are imposed by the source size, the camera field of view (FOV), and whether the UUT surface can reflect light emitted from the source. If a deflectometry setup can capture light reflected by the UUT, it can measure the local slopes of the UUT surface (Graves et al., 2019). Combined with precise system calibration data, slopes may be integrated in post-processing to reconstruct surface maps with accuracy comparable to that produced by interferometry (Martin et al., 2018).

In a typical deflectometry configuration, a high-resolution camera possessing a well-defined entrance pupil location, referred to as $p(x,y,z)$, focuses onto the UUT surface. Camera pixels are mapped to the UUT surface, which is represented by discrete 'mirror pixels', referred to as $u(x,y,z)$. Ideally, the source for a deflectometry setup, referred to as $s(x,y,z)$, has high stability, repeatability, and signal power, which provide the test system with a high signal-to-noise ratio (SNR). At a single camera pixel that successfully captures light reflected from the UUT surface, the precise location on the source that illuminated the camera pixel is determined. Using the determined ray start location at the source, the end location at the camera, and the intercept location at the mirror pixel, the local slope at the mirror pixel can be calculated. This process is extended to all camera pixels to measure the local slopes at all mirror pixels on the UUT in orthogonal directions, referred to as $S_X(x,y,z)$ and $S_Y(x,y,z)$, representing the x and y slopes respectively. These slope maps are typically integrated with a zonal integration method such as Southwell integration (Southwell, 1980) or a modal integration such as using a gradient Chebyshev polynomial set (Aftab et al., 2018), resulting in a reconstructed surface map. Figure 1 demonstrates a standard deflectometry setup and the model used for local slope calculation.

2.2 *Infrared deflectometry and existing tungsten wire paradigm*

Infrared deflectometry extends deflectometry to measuring diffuse rough optics, which are challenging to measure using traditional techniques. Because a wide range of materials and surfaces do not specularly reflect visible light, infrared deflectometry is an important metrology tool. This is particularly true during the grinding phase of mirror fabrication, where a rough grit is used to rapidly grind the UUT down to the final desired surface shape. During this period, the root-mean-square (RMS) surface roughness will typically drop from 100 μm to 1 μm as smaller grit sizes are used. For such rough surfaces, visible light is scattered and thus visible spectrum metrology tools are inapplicable; however, infrared deflectometry has been applied during this phase successfully for several mirror fabrication projects, including the Daniel K. Inouye Solar Telescope (DKIST) primary mirror (Oh et al., 2016).

Traditionally, a heated tungsten ribbon acts as the infrared source, serving as a pseudo-blackbody element. Coupled with a LWIR camera, which is sensitive in the 7-14 μm range, this source allows for testing 1 μm to ~25 μm RMS rough surfaces (Su et al., 2013; Kim et al., 2015, 2016; Oh et al., 2016). This test setup has been successfully used to measure a variety of rough, non-specularly reflecting surfaces and was able to achieve high accuracy surface reconstruction. It should be noted that other dynamic thermal pattern generators, including a scanning infrared laser and a resistor array, have successfully been used as sources for infrared deflectometry. However, a heated scanning ribbon still serves as the most common source for testing large diffuse optics (Höfer et al., 2016).

2.3 *Compact, infrared pseudo-blackbody emitter*

While a tungsten ribbon source can be engineered for higher signal power or larger size, its design is particularly vulnerable to thermal fatigue and suffers from several inherent limitations. Cyclically heated tungsten evaporates and degrades with use over time, leading to a potentially non-uniform emission profile across the surface. This is coupled with the nature

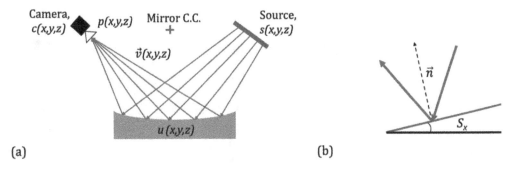

(a)   (b)

Figure 1.   In a standard deflectometry system (a), light is emitted by a source, $s(x,y,z)$. Some of the emitted light will deflect from the UUT surface, $u(x,y,z)$, and be successfully captured by the camera, $c(x,y,z)$, with an entrance pupil location, $p(x,y,z)$. Represented by the vector $\vec{v}(x,y,z)$, a deflected light ray follows the law of reflection and is reflected by a local area on the UUT surface, whose normal vector is given by $\vec{n}$. Knowing the ray start, intercept, and end location, the local slope of the UUT, given by S_X, can be determined (b).

5

of a thin ribbon to possibly experience any of its lower bending modes during testing, further deviating the emission area from the theoretical ideal rectangular area. Additionally, driving higher input current through the ribbon only accelerates material degradation and also shifts the output spectrum away from desirable longer wavelengths. Finally, the heated tungsten ribbon invariably creates a temperature gradient in the surrounding air due to local convective boundary conditions. For larger ribbons and longer source operation time, growing pockets of locally heated air may generate thermal noise that blurs the infrared source signal.

We propose an alternative source called a Long-wave Infrared Time-modulated Integrating cavity Source (LITMIS) to address these limitations. At the core of the LITMIS design is modular usage of EMIRS200 Series emitters, which are available commercially off-the-shelf from Axetris AG. The packaged emitters, henceforth referred to as 'caps', are typically used in nondispersive infrared (NDIR) and photoacoustic gas spectroscopy and have a usable lifetime of 10,000 hours (Esfahani & Covington, 2017; Wilson et al., 2019). Although the emissive spectral distribution is centered at 4 μm at the operational temperature (550°C), radiant flux across the desired 7-12 μm band of interest is also produced, as described by Planck's Law. Caps are fabricated by electroplating Platinum Black, a platinum powder, onto a thin membrane which floats on a silicon substrate (Hessler et al., 2004; Axetris AG, 2014, 2019). Similar to the tungsten ribbon, a cap is operated by driving an electric current through the target material to induce resistive heating and emit blackbody radiation. Two desirable source properties result from the cap material design and geometry:

- High emissivity: when electroplated onto a substrate, Platinum Black forms dendritic structures on the scale of < 50 nm. The feature scale and high porosity of the structure result in high absorptivity and high emissivity ($\varepsilon > 0.9$) characteristics for the coated membrane, unlike that of smooth metallic surfaces. For example, the typical emissivity of tungsten is $\varepsilon < 0.1$ at 1000K (Verret & Ramanathan, 1978).
- Low thermal capacitance: this property, also called thermal mass, refers to a body's ability to store thermal energy. Here, the combined resistive network of an exterior Platinum Black coating, Tantalum Oxide, and a SiN membrane is only several microns thick (Hessler et al., 2004), as compared to a typical ~30 μm thick tungsten ribbon. Because the ultra-low mass is suspended in air by a silicon substrate, the thin emissive network can heat to and cool from its steady-state temperature in less than 1 second (Hessler et al., 2004).

Temporal modulation emerges as a viable operational mode of the new source. Testing reveals that a cap can achieve an 80% contrast ratio at 1 Hz, which enables in-situ background noise images to be taken during testing. Therefore, at a given image capture, a pair of 'signal' and 'background' shots can be successively captured for noise subtraction in post-processing.

 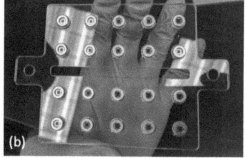

Figure 2. The Axetris EMIRS200 TO39 source features a resistive membrane that, when current is applied, acts as a pseudo-blackbody source (a). The emitters can be connected in series or in parallel, much like a group of resistors, to create a patterned source (b).

This option is not practical for the tungsten ribbon source because its relatively higher thermal mass evinces a slower thermal transient response. For any thermal source, noise will accumulate in the environment as it warms up and approaches its steady-state operational temperature.

2.4 Long-wave infrared time-modulated integrating cavity source

The Long-wave Infrared Time-modulated Integrating cavity Source (LITMIS) leverages the EMIRS200 elements as flexibly placed, highly time-responsive input nodes for radiative emission. Nodes radiate LWIR spectra into the integrating cavity, where the light is scattered and achieves a uniform, non-directional emission when exiting the rectangular output slit. Cavity length, interior wall roughness, and exit slit geometry are all optimized to minimize reflection losses. Furthermore, interior walls may be protected with high reflectivity coatings, like silver or gold, and the number of input caps may be increased to scale power output.

For direct comparison with the ribbon source modality, an integrating cavity was designed and machined to match the geometrical and radiometric properties of an existing, functional tungsten rectangular ribbon source used for deflectometry (Su et al., 2011). These properties include source surface area and source radiance. The cavity was designed with 20 input 'cap' sources, operating at approximately 70% maximum power for safety and in order to achieve a 1 Hz flicker rate. The cavity itself was optimized to achieve both spatial uniformity over a rectangular exit which featured a pseudo-Lambertian emission angle, while the interior of the cavity was a box shape made of bare aluminum with a surface roughness of 3.4 μm RMS. The system was modeled in LightTools, a non-sequential ray tracing simulation software, and the location of the heating elements, as well as interior cavity dimensions and surface roughness, were optimized to achieve a uniform power output across the exit slit while maintaining non-directional output over approximately 2π steradians. The output was simulated at the slit, where uniform power was the goal. The near field irradiance pattern, as well as the final optimized box design, are shown below in Figure 3.

3 LITMIS EXPERIMENTAL MEASUREMENT SETUP AND RESULTS

3.1 Deflectometry hardware, UUT, and source configuration

An infrared deflectometry system was configured such that all hardware was common, except the source, which was changed during experimental measurements. The Thermal-Eye 3500AS LWIR camera featured a ~1 – 2 m variable focal length germanium lens, and its detector was

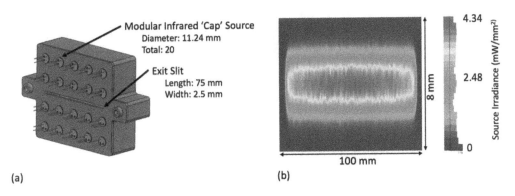

(a)   (b)

Figure 3.   The final optimized LITMIS assembly was modeled in SolidWorks (a). 20 infrared emitters point inside a diffusing cavity which contains a single exit slit, machined from a thin aluminum plate. The optimized design was also modeled in LightTools, where the irradiance at the surface of the box was simulated to assure high uniformity across the exit slit (b).

a 160×120 microbolometer array with 320×240 super-resolution output. Exposure, gain, and level settings were adjusted to prevent output saturation and were held constant between all tests. A fixed optical mount was situated approximately one meter from the camera and allowed for repeatable UUT placement. To compare properties between source modalities, a scanning platform was utilized with a mounting interface to interchange LWIR sources. A motorized lead screw stage moved source assemblies in the vertical direction with an absolute positional accuracy of ± 0.005 mm. For ease of comparison, sources shared both identical slit dimensions (75 mm $\times$ 2.5 mm) and radiant exitance planes. Figure 4 demonstrates the camera and source setup for the test system.

UUTs included a 2-inch diameter rough ground glass referred to as $Glass^{1500}$ and a bare aluminum flat referred to as Al_{Room}. UUTs were selected for measurement due to their diffuse nature, making thermal infrared deflectometry an ideal metrology method. When heated to 150°C, Al_{Room} flat is referred to as Al_{150}, and is done so as a challenge case because it generates variable thermal background noise, which typically degrades the SNR of a test. This is a common scenario, as several optics, such as solar collectors or even the DKIST primary mirror, will operate under thermal load; thus, the Al_{150} test case is highly relevant. The surface roughness of both optics was measured using a Zygo NewView 8300 Interference Microscope. The ground glass surface featured a surface roughness of 127.89 nm RMS while the bare aluminum surface roughness was 102.53 nm RMS over a small 834×834 μm square area over each optic.

The LWIR line source was implemented by running direct current (2.2 A, 2.1 W) across a thin 32 μm rectangular tungsten ribbon. Transient thermal noise from the wire, such as the local heating of air, was reduced by taking measurements only after the ribbon reached a thermal steady-state condition, which was achieved after approximately 20 minutes.

The LITMIS source was implemented by applying a (0.29 A, 7 W) load to a circuit consisting of 20 emitters. Pointed into the enclosure, the rectangular emitter array was operated by binary power cycling with a digitally controlled relay. Enclosure walls were machined from bare Al 6061-T6 and characterized to 3.4 μm RMS by a Zygo NewView 8300 Interference Microscope. Lastly, to minimize latent thermal radiation from the cavity interior, the LITMIS source was cooled to 0°C prior to each test. Figure 5 displays the integrating cavity after assembly.

3.2 *Measurements and results for source geometry and temporal stability*

Geometrical profile measurements were conducted by recording an image focused at each source with the LWIR camera. The camera software was adjusted so the maximum 8-bit signal count was a discrete value just below 255. Source profiles were captured and compared against

Figure 4.   A traditional tungsten ribbon scanning source is mapped to Thermal-Eye 3500AS camera. As a requirement for deflectometry measurement, the camera is focused on the surface of the UUT. LWIR radiation leaves the source, of which, some rays are intercepted by the detector. For ground optics, higher SNR is required to precisely, or at all, determine the surface slope at a given mirror pixel. After UUT measurement with the ribbon, it is unmounted and replaced by the LITMIS source.

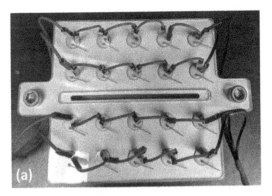

Figure 5. A frontal view of the assembled LITMIS architecture is shown in (a). Four parallel chains of elements are connected to operate as a single electrical circuit. While radiative thermal input is intended towards the interior of the box, excess thermal radiation is emitted by the heated copper body of each cap. Aluminum-covered shielding is mounted to the LITMIS façade to deflect it, as shown in (b). By design, LWIR radiation does not propagate directly from a heat source to the object of measurement. Rather, it is diffused by internal scattering and may only escape through a limited window. With this mechanism, input power can be readily scaled by simply adding more emitters within the cavity. In contrast, the available input power from a direct thermal source is limited by the area of its profile geometry.

to an assumed ideal rectangular emission profile. As seen in Figure 6, the LITMIS source maintains a flatter signal power across the horizontal pixel band and drops off sharply at the edges.

To observe the temporal stability, a measurement was performed by focusing the camera on each source, which was turned on and recorded for 30 minutes, separately. Temporal measurement results illustrate the high stability and uniform slit emission

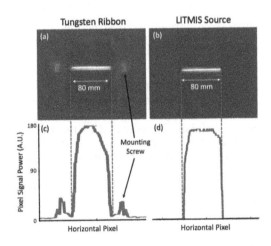

Figure 6. The images of the tungsten ribbon (a) as well as the LITMIS slit (b) sources were captured using Thermal-Eye 3500AS camera through-focused onto the source through a flat mirror. Observing the camera signal power histograms for the tungsten ribbon (c) and the LITMIS slit (d) sources, it is seen that the average power is similar, but the source profile geometries are quite different, where both should ideally form a flat top rectangular shape with roll-on and roll-off at the edges. This is due to the convolution of the rectangular source with the circular camera pupil.

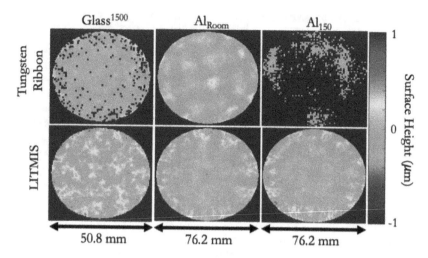

Figure 7. Using both a traditional tungsten ribbon source (top row) and the LITMIS source (bottom row), infrared deflectometry measurements were taken and the surface reconstructed for the Glass^{1500} optic (left column), the Al$_{Room}$ optic (middle column), and the Al$_{150}$ optic (right column). For all maps, Standard Zernike terms 1:37 were removed to observe the surface mid-to-high spatial frequency topology. In the Glass^{1500} case, some points in the aperture could not be sampled due to insufficient SNR. In the Al$_{150}$ case, a majority of the surface topology was unavailable due to competitive thermal noise. With greater immunity from thermal noise, the LITMIS source obtained sufficient slope data to reconstruct all three samples.

geometry of the LITMIS source. Stability was most pronounced in the signal peak-to-valley fluctuation comparison, which was measured as 11 signal counts for the tungsten ribbon source and 1.82 for the LITMIS source.

3.3 *Measurements and results for UUT surface reconstruction and repeatability*

To determine the comparative surface reconstruction repeatability, the Glass^{1500}, Al$_{Room}$, and Al$_{150}$ optics were measured using both tungsten ribbon and LITMIS sources.

The reconstructed surface topology maps of all 3 UUT cases are shown in Figure 7. Both the LITMIS slit source and tungsten ribbon source were successfully used to test the Glass^{1500} and Al$_{Room}$ optics. However, the standard deviation of the signal power across the five repeat measurements performed for every optic for each source was slightly larger for the tungsten ribbon as compared to the LITMIS source. Despite this, the LITMIS source was better able to reduce noise for all cases, which directly impacts the SNR of both test methods. Overall, the LITMIS source achieved a 2–5 times larger SNR for the Glass^{1500} and Al$_{Room}$. The Al$_{150}$ sample could not be measured using the ribbon source due to high thermal noise generated by the UUT, while the LITMIS source provided sufficient slope data for surface reconstruction.

4   CONCLUSIONS

In infrared deflectometry, any uncertainty in the spatial and temporal behavior of a source directly negatively impacts the reconstruction accuracy and uncertainty. Additionally, time-varying, thermal background noise is common in most test environments and degrades the effectiveness of LWIR sources by decreasing the signal-to-noise ratio. We have instead created an integrating cavity source, which emits long-wave infrared light uniformly from a defined exit slit, which we call the LITMIS source. A demonstration infrared deflectometry system using the new source successfully tested a ground glass sample and aluminum blank, as well as a previously unmeasurable aluminum blank under thermal load. In all cases, the source exhibited excellent repeatability and significantly improved the SNR of the test, as compared to testing using a traditional tungsten ribbon. The authors encourage readers to examine further detailed test results and analysis in the paper, *"High-contrast thermal deflectometry using long-wave infrared time-modulated integrating cavity."* (Graves et al., in prep.)

The scalability and flexibility of the new source architecture leave much to explore. Beyond simply engineering the source design to achieve high signal output power, temporal behavior may be exploited to reconstruct shapes of challenging optics. As temporal modulation adequately inoculates measurement from local thermal fluctuations in active workpieces, the creation of an on-machine deflectometry system also edges towards the realm of possibility. More excitingly, these building blocks may be applied towards a sinusoidally modulated infrared source, which may allow phase-shifting infrared deflectometry to become viable in optical metrology.

ACKNOWLEDGMENTS

The authors would like to acknowledge the II-VI Foundation Block-Gift Program for helping support general deflectometry research in the LOFT group, making this research possible. Also, this work was made possible in part by the Technology Research Initiative Fund Optics/Imaging Program and the Korea Basic Science Institute Foundation.

REFERENCES

Aftab, M., Burge, J.H., Smith, G.A., Graves, L.R., Oh, C.J. and Kim, D.W. 2018. 'Chebyshev gradient polynomials for high resolution surface and wavefront reconstruction', 1074211(September 2018), p. 40. doi:10.1117/12.2320804.

Axetris AG 2014. 'IRS Frequently Asked Questions Rev.A', pp. 2–5.

Axetris AG 2019. 'Infrared Sources Brochure (English)', pp. 42–75. doi:10.1201/9780203750834-3

Dubin, M.B., Su, P. and Burge, J.H. 2009. 'Fizeau interferometer with spherical reference and CGH correction for measuring large convex aspheres', *Optical Manufacturing and Testing VIII*, 7426 (May), p. 74260S. doi:10.1117/12.829053.

Esfahani, S. and Covington, J.A. 2017. 'Low Cost Optical Electronic Nose for Biomedical Applications', *Proceedings*, 1(10), p. 589. doi:10.3390/proceedings1040589

Fang, F.Z., Zhang, X.D., Weckenmann, A., Zhang, G.X. and Evans, C. 2013. 'Manufacturing and measurement of freeform optics', *CIRP Annals - Manufacturing Technology*. CIRP, 62(2), pp. 823–846. doi:10.1016/j.cirp.2013.05.003.

Graves, L.R., Quach, H., Koshel, J.R., Oh, C.J., Kim, D.W. (in prep). 'High contrast thermal deflectometry using long-wave infrared time-modulated integrating cavity.'

Graves, L.R., Quach, H., Choi, H. and Kim, D.W. 2019. 'Infinite deflectometry enabling 2π-steradian measurement range', *Optics Express*, 27(5), p. 7602. doi:10.1364/oe.27.007602.

Hessler, T., Dubochet, O., Forster, M. and Merschdorf, M. 2004. 'Micro-machined, electrically modulated thermal infrared source with black body characteristic pyroelectric detectors or', (November 2015).

Höfer, S., Burke, J. and Heizmann, M. 2016. 'Infrared deflectometry for the inspection of diffusely specular surfaces', *Advanced Optical Technologies*, 5(5–6), pp. 377–387. doi:10.1515/aot-2016-0051.

Kim, D.W., Su, T., Su, P., Oh, C., Graves, L. and Burge, J. 2015. 'Accurate and rapid IR metrology for the manufacture of freeform optics', *SPIE Newsroom*, pp. 9–11. doi:10.1117/2.1201506.006015.

Kim, D.W., Oh, C., Lowman, A., Smith, G.A., Aftab, M. and Burge, J.H. 2016. 'Manufacturing of super-polished large aspheric/freeform optics', *Advances in Optical and Mechanical Technologies for Telescopes and Instrumentation II*, 9912, p. 99120F. doi:10.1117/12.2232237.

Lowman, A.E., Yoo, H., Smith, G.A., Oh, C.J. and Dubin, M. 2018. 'Improvements in the scanning long-wave optical test system', 1074216(September 2018), p. 48. doi:10.1117/12.2321265.

Martin, H.M. *et al.* 2018. 'Manufacture of primary mirror segments for the Giant Magellan Telescope', (May), p. 30. doi:10.1117/12.2312935.

Oh, C.J. *et al.* 2016. 'Fabrication and testing of 4.2m off-axis aspheric primary mirror of Daniel K. Inouye Solar Telescope', 9912, p. 99120O. doi:10.1117/12.2229324.

Southwell, W.H. 1980. 'Wave-front estimation from wave-front slope measurements', *Journal of the Optical Society of America*, 70(8), p. 998. doi:10.1364/josa.70.000998.

Su, T., Park, W.H., Parks, R.E., Su, P. and Burge, J.H. 2011. 'Scanning Long-wave Optical Test System: a new ground optical surface slope test system', *Optical Manufacturing and Testing IX*, 8126(May), p. 81260E. doi:10.1117/12.892666.

Su, T., Wang, S., Parks, R.E., Su, P. and Burge, J.H. 2013. 'Measuring rough optical surfaces using scanning long-wave optical test system 1 Principle and implementation', *Applied Optics*, 52(29), p. 7117. doi:10.1364/ao.52.007117.

Verret, D.P. and Ramanathan, K.G. 1978. 'Total hemispherical emissivity of tungsten', *Journal of the Optical Society of America*, 68(9), p. 1167. doi:10.1364/JOSA.68.001167.

Wilson, D., Phair, J. and Lengden, M. 2019. 'Performance Analysis of a Novel Pyroelectric Device for Non-Dispersive Infra-Red CO_2 Detection', *IEEE Sensors Journal*, 1748(c), pp. 1–1. doi:10.1109/jsen.2019.2911737.

Advances in Optoelectronic Technology and Industry Development – Jose & Ferreira (eds)
© 2020 Taylor & Francis Group, London, ISBN 978-0-367-24634-1

Adaptive learning rate and target re-detection for object tracking based on correlation filter

Jianhong Xiang & Pengyu Shen
College of Information and Communication Engineering, Harbin Engineering University, Harbin, China

ABSTRACT: In this paper, two problems about the update rate and long-term tracking in target tracking model are discussed. Traditional correlation filter tracker only uses a fixed rate mechanism, so the target update rate is fixed. In this paper, we improved it so that it can adjust the update rate adaptively according to the similarity between different image sequences and first frame. Besides, in order to deal with more challenging scenarios and long-term tracking targets, we add a re-detection mechanism to the tracker. This method overcomes the limitation of the traditional correlation filter tracker using fixed update rate by studying the similarity between the frames of the image, and can adaptively change the update rate of the model. A large number of experimental results show the superiority of our improved tracker in accuracy and success rate.

1 INTRODUCTION

Object tracking is the core of computer vision and has been widely used in surveillance, human-computer interaction, robotics and other fields Wibowo,2018).In recent years, the tracker based on kernel correlation filter has excellent performance and it can maintain high frame rate with high accuracy, and the current mainstream benchmark rankings are very high. However, most of the kernel correlation filtering algorithms adopt a fixed learning rate, which cannot adapt to the change of tracking environment. Moreover, most short-term tracking has no target re-detection mechanism, and cannot deal with the long-term tracking problem well.

The proposed method is closely related to the correlation filter-based tracker which applies the correlation filter in traditional signal processing technology to tracking applications. Visual tracking has been studied extensively with numerous applications (Smeulders, 2014). In 2010, Bolme et al. (2010) proposed a Minimum Output Sum of Squared Error (MOSSE) filter. It is the first time to apply the correlation filter to the tracking field. The MOSSE is simple in calculation and can track the target quickly, but it cannot guarantee accurate tracking when the appearance of the target changes. In 2014, Li et al. fused the features of color-naming (CN) (van de Weijer, 2009) and histogram of orientation gradients (HOG) and predicted the scale change of target by using scale pool. Then, the Scale Adaptive with Multiple Features Tracker (SAMF) was proposed (Li, 2014). SAMF improved the discriminant ability of target on the basis of High-speed tracking with kernelized correlation filter (Henriques, 2015) and solved the scale change problem to a certain extent. It has good tracking effect, but it still needs to be improved for fast moving target and partly occluded target tracking. In 2015, Ma Chao et al. proposed LCT algorithm which takes into account the context and scale transformation of the target (Ma, 2015). Based on the correlation filtering, a random fern-based re-detector is combined to further improve its long-term tracking effect. The traditional tracking framework has boundary effect. Mueller et al. (2017) proposed a Context-Aware (CA) framework to solve the boundary effect caused by cyclic sampling which can be integrated with many classical CF trackers. Tracking-by-detection trackers have attracted wide attention due to their high performance and efficiency (Kalal, 2012).

In this paper, we mainly study single target tracking problems. We consider the problems mentioned above and propose a novel Adaptive learning rate and re-detection tracker (ALRD). The proposed method overcomes the limitation of the traditional correlation filter tracker using fixed update rate by studying the similarity between each frame of the image and the first frame to adaptively change the update rate of the model. Meanwhile, in order to deal with more challenging scenarios and long-term tracking targets, we have added a re-detection mechanism to the tracker.

The main contributions of this paper are summarized below: (1) The fixed update rate of the tracker based on correlation filter is improved so that it can adapt the update rate on the basis of the different image sequences; (2) A re-detection mechanism is added to the tracker so that the tracker can track the target for a long time; (3) A large number of experiments have been carried out to compare the improved tracking algorithm and some short-term tracking algorithm with our proposed method, which includes adaptive update rate and increasing re-detection mechanism to cope with long-term tracking, and the results show the better performance of the proposed tracker in terms of accuracy and success rate.

2   RELATED WORK

The tracker proposed in this paper is based on SAMF tracking of CA framework. In the following passage, we use SAMF(CA) to represent based on SAMF tracking of CA framework. We introduce the differential hashing algorithm in order to solve the update rate is a fixed value in target tracking this problem.

2.1   *SAMF algorithm based on CA framework*

The traditional CF tracker uses discriminant learning as its core. Sampling mode of SAMF algorithm is the same as that of High-speed tracking with kernelized correlation filter. All training samples are obtained by cyclic displacement of target samples. Suppose that we have a one-dimensional data $x = [x_1, x_2, \ldots, x_n]$, a cyclic shift of x is $\mathbf{P}x = [x_n, x_1, x_2, \ldots, x_{n-1}]$. Finally, all cyclic matrices can be expressed as below:

$$\mathbf{X} = \mathbf{F}^H \, \mathrm{diag}(\mathbf{F}\mathbf{x})\mathbf{F} \tag{1}$$

where $\mathbf{F}$ is the DFT matrix and $\mathbf{F}^H$ is the Hermitian transpose of $\mathbf{F}$. The cyclic structure of this matrix helps to effectively solve the problem of regression in the Fourier domain ridge.

$$\min_{\mathbf{w}} \|\mathbf{A_0}\mathbf{w} - y\|_2^2 + \lambda_1 \|\mathbf{w}\|_2^2 \tag{2}$$

Here, the learning correlation filter is represented by a vector $\mathbf{w}$. $\mathbf{A_0}$ represents the cyclic shift matrix, and the regression target is y. λ_1 is a regularization factor parameter.

Because the surrounding environment of the tracked object has a great impact on the tracking performance of the tracker, for example, there is a lot of background chaos around the tracked object. In order to overcome this situation, a context-based CF tracking framework (CA framework) is introduced. In the learning stage, context blocks are added around the tracking target by adding context information to the filter. The formula can be expressed as follows:

$$\min_{\mathbf{w}} \|\mathbf{A_0}\mathbf{w} - y\|_2^2 + \lambda_1 \|\mathbf{w}\|_2^2 + \lambda_2 \sum_{i=1}^{k} \|\mathbf{A}_i\mathbf{w}\|_2^2 \tag{3}$$

where $\mathbf{A}_i$ is the cyclic matrix corresponding to context patch. λ_2 is a context patch regression parameter. SAMF (CA) is an algorithm based on Equation 3:

$$f_p(\mathbf{w}, \mathbf{B}) = \|\mathbf{B}\mathbf{w} - \bar{\mathbf{y}}\|_2^2 + \lambda_1 \|\mathbf{w}\|_2^2 \tag{4}$$

where

$$\mathbf{B}=\begin{bmatrix} \mathbf{A_0} \\ \sqrt{\lambda_2}\mathbf{A_1} \\ \vdots \\ \sqrt{\lambda_2}\mathbf{A}_k \end{bmatrix} \text{ and } \bar{y}=\begin{bmatrix} y \\ 0 \\ \vdots \\ 0 \end{bmatrix}. \tag{5}$$

Because $f_p(\mathbf{w}, \mathbf{B})$ is convex, it can be minimized by setting the gradient to zero, yielding:

$$w = (\mathbf{B}^T\mathbf{B} + \lambda_1\mathbf{I})^{-1}\mathbf{B}^T\bar{y} \tag{5}$$

We linearly combine the new filter with the old one as below:

$$\bar{T} = \theta T_{new} + (1 - \theta)\bar{T} \tag{6}$$

where $\bar{T}$ is the template to be updated.

2.2 Different hash algorithm

Perceptual hashing algorithm is the general term of a class of algorithms, including average hash, perceived hash and different hash. According to the characteristics of perceptual hashing, it can map objects with large amount of data to a series of bits with smaller growth. When it applies to image matching, the image generates a fingerprint (string format). The more similar the fingerprints of the two pictures are, the more similar the two pictures are. Average hash: It's faster, but the results are not very accurate. Perceptual Hash: It is more accurate, but slower in speed. Differential hashing: High accuracy, and very fast. Therefore, this paper uses differential hashing to measure similarity. The specific steps are as in Algorithm 1.

Algorithm 1: differential hashing algorithm

Input: image
Output: different hash value
1:Reducing image size
  Reduce the size of the picture to 8x8 and there are 64 pixels in total. The purpose of this step is to remove the differences in the size and proportion of various pictures, and only retain basic information such as structure, light and shade.
2:Conversion to grayscale images
  Difference hashing is achieved by calculating the color intensity difference between adjacent pixels, so the reduced image is converted to 64-level gray image.
3:Calculating the average gray level
  Calculate the average gray level of all the pixels in the picture
4:Comparing the gray levels of the pixels
  Compare the gray level of each pixel with the average value. If the average value is greater than or equal to 1, the average value is less than 0.
5:Calculating hash values
  A 64-bit binary integer is formed by combining the results of the previous step together, which is the fingerprint of this picture and the final different hash value.

---

3 TRACKING COMPONENTS

3.1 Adaptive template updating

In the actual updating process, if the model updates too slowly to keep up with the change of target features, it cannot meet the accuracy. Blindly increasing the updating rate will lead to errors, noise and other problems. And then, these will cause the model drift. Therefore, ALRD formulates the update mechanism from the similarity between each frame of the image and the first frame. It is shown in Figure 1.

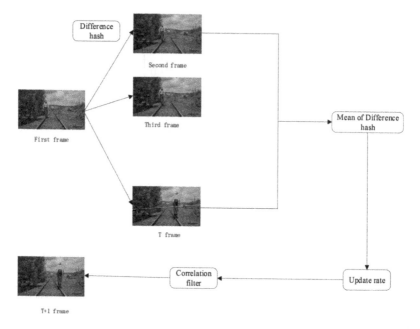

Figure 1.   Adaptive updating framework.

In order to adapt to the similarity between frames, an update rate related to image similarity is constructed. We considering the relationship between each frame and the first frame of target tracking image. If the similarity between the current frame and the first frame is high and the change of tracking target is small, the correlation filter can be updated with a smaller update rate. If the similarity between the current frame and the first frame is low, the change of target may be large, and the correlation filter can be updated with a larger update rate. The relationship between update rate and image frame similarity is proposed as follows:

$$\theta = \frac{1}{1 + (5 + m_{dHash})^3} \tag{7}$$

where θ is the update rate and m_{dHash} is the mean of the difference hash value. According to Equation 7, we can realize adaptive update rate.

3.2   Target re-detection mechanism

SAMF (CA) tracking algorithm is a short-term tracking algorithm. Therefore, combined with the re-detection mechanism of LCT tracking algorithm, the re-detection based on confidence filter and support vector machine are added to SAMF (CA) tracking algorithm to improve the long-term tracking effect. The location of the target is determined by a short-term tracker and a confidence filter R. The maximum response value is calculated around the target position by confidence filter R_t as tracking confidence. When the confidence level is lower than the set threshold, the tracking target is considered to have an error. The re-detector is activated and the tracking target position is updated according to the re-detector results.

When the maximum response value of the confidence filter is greater than the set threshold value, the result of short-time filtering tracking is considered to be reliable enough. At this time, the result of short-time filtering is used to locate the target, and the confidence filter and SVM classifier are updated. When the maximum response value of the tracking target obtained by the confidence filter is less than the set threshold value, it is considered that the tracking result of the short-time tracker is wrong and the re-detection mechanism is called to locate the target according to the re-detection tracking result.

16

$$P_F = \begin{cases} P_{CF}, & R_t > T \\ P_{SVM}, & R_t < T \end{cases} \tag{8}$$

where P_F is the final target location and P_{CF} is the target location determined by adaptive update rate filter. P_{SVM} is the target location based on SVM re-detection mechanism. An overview of the proposed method is summarized in Algorithm 2.

Algorithm 2 ALRD tracking algorithm

Input: Initial target bounding box x_0
Output: Central Point Position of Target in Each Frame, Adaptive update rate filter, Confidence Filter and SVM detector D_{svm}.
1: **repeat**
2: Crop out the image region in frame t account to P_{t-1} and extract the features.
3: Compute the correlation map y_t using P_{t-1} and R_a to estimate the position P_t.
4: Calculate the difference hash value d_{Hash} between the $t-1$ frame and the t frame.
5: Calculate the mean of the difference hash value m_{dHash} and get the update rate by Equation 7.
6: **if** max $(y_t) < T$ **then**
7: Use detect D_{svm} to perform re-detection and calculate confidence score y_i'.
8: **if** $y_i' > 0$ **then** Use detect D_{svm} to estimate the position P_t.
9: Update R_a.
10: **if** max $(y_t) > T$ **then** update R_t.
11: Update D_{svm}.
12: **until** end of video sequence.

---

4 EXPERIMENT

To verify the effectiveness of our proposed tracker, we compared it with four CF trackers. To evaluate, we used the popular target tracking benchmark OTB-100 (Wu, 2015).

4.1 Implementation details

Since the re-detection mechanism is compared with the fixed threshold, different threshold selection will affect the final experimental results. If the threshold is too large, the re-detection mechanism will be started frequently. Then it will reduce the real-time performance. If the threshold is too small, the re-detection mechanism will not be able to enter in the case of target tracking errors. Therefore, tracking accuracy and success under different thresholds are tested, as shown in Table 1. The regularization factor of Equation 3 is set to $\lambda_1 = 10^{-4}$. The additional regularization factor is set to $\lambda_2 = 0.4$. We set the number of context patches k to 4. The Gaussian kernel width σ is set to 1. The size of the search window is set to 2 times the target size. The proposed tracking algorithm is implemented in MATLAB software on an Intel I5-8300H2.3 GHz CPU with 8 GB RAM. The running speed of the ALRD algorithm is 4.74 FPS.

According to the comparison of different thresholds, it can be concluded that the target tracking success rate and accuracy rate are the best at threshold $T = 0.7$.

4.2 Analyses of ALRD

OPE is a common evaluation criterion for tracking algorithms. This paper tests all 100 video sequences in OTB100. The ALRD algorithm has a success rate of 4.96% higher than SAMF

Table1. Comparison of success rate and precision rate under different thresholds.

Threshold	0.4	0.5	0.6	0.7	0.8	0.9
Precision	0.811	0.812	0.812	0.818	0.807	0.806
Success	0.713	0.714	0.714	0.719	0.710	0.711

(CA) algorithm at 0.5 threshold. The accuracy in 20 pixels is 4.47% higher than that of SAMF (CA) algorithm. The results are shown in Figure 3.

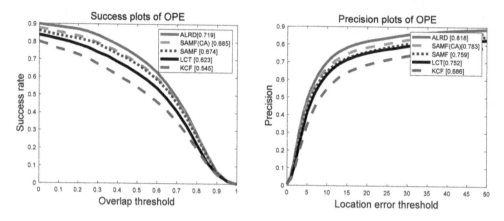

Figure 3. Success and accuracy diagrams of OPE tracking algorithms.

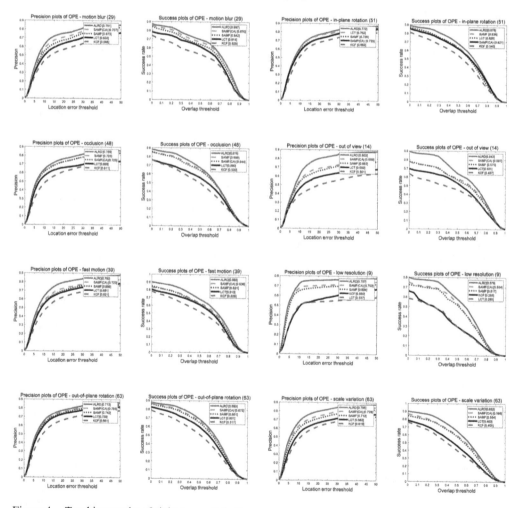

Figure 4. Tracking results of eight types of challenge sequences.

4.3 Evaluation on ALRD

To illustrate the advantages of the proposed algorithm in the current tracking field, eight types of challenges in the OTB100 dataset are selected and compared with four other tracking methods. The results are shown in Figure 4. Among existing methods, the SAMF(CA) method performs well with overall success in motion blur (67.0%), fast motion (63.6%), out-of-plane-rotation (67.2%), low resolution (53.4%), scale variation (59.8%) and out-of-view (58.7%), while the ALRD algorithm achieves success rates of 69.7%, 68.3%, 69.3%, 57.9%, 65.2% and 64.3%, respectively. The SAMF method performs well in in-plane-rotation (63.6%) and occlusion (65.6%), while the ALRD algorithm achieves success rates of 67.8% and 67.5%. In terms of scale variation, the SCM method achieves the success rate of 51.8% while the LCT algorithm performs well with a success rate of 55.8%. Especially in the case of out-of-view and fast motion, we can see the advantages of the ALRD algorithm through the comparison above.

5 CONCLUSION

This paper improves SAMF (CA) from two aspects: template dynamic update rate and re-detection mechanism. Firstly, the similarity between adjacent frames is used to dynamically adjust the update rate. Secondly, to improve the ability of SAMF (CA) long-term tracking, a LCT framework strategy is proposed, and a confidence filter and a SVM-based classifier re-detection mechanism are added. Through detailed analysis of the proposed algorithm and experimental comparison with several algorithms in OTB100 database, the tracking accuracy and success rate of this algorithm are improved compared with SAMF (CA) and the tracking performance of the proposed algorithm is better than that of other algorithms under various challenges of target tracking.

ACKNOWLEDGMENTS

This paper is supported by the National Key Laboratory of Communication Anti-jamming Technology (614210202030217). This paper is also funded by the International Exchange Program of Harbin Engineering University for Innovation-oriented Talent Cultivation.

REFERENCES

Bolme, D.S., Beveridge, J.R., Draper, B.A., & Lui, Y.M. 2010. *Visual object tracking using adaptive correlation filters. Computer Vision & Pattern Recognition*:2544–2550.
Henriques, J.F., Caseiro, R., Martins, P., & Batista, J. 2015. High-speed tracking with kernelized correlation filters. *IEEE Transactions on Pattern Analysis and Machine Intelligence* 37(3):583–596.
Kalal, Z., Mikolajczyk, K., & Matas, J. 2012. Tracking-learning-detection. *IEEE Trans Pattern Anal Mach Intell* 34(7):1409–1422.
Li, Y., & Zhu, J. 2014. A Scale Adaptive Kernel Correlation Filter Tracker with Feature Integration. *European Conference on Computer Vision*. Cham: Springer.
Ma, C., Yang, X., Zhang, N.C., & Yang, M.H. 2015. Long-term correlation tracking. *2015 IEEE Conference on Computer Vision and Pattern Recognition (CVPR)*. IEEE Computer Society.
Mueller, M., Smith, N., & Ghanem, B. 2017. Context-Aware Correlation Filter Tracking. *2017 IEEE Conference on Computer Vision and Pattern Recognition (CVPR)*. IEEE Computer Society.
Smeulders, A.W.M., Chu, D.M., Cucchiara, R., Calderara, S., Dehghan, A., & Shah, M. 2014. Visual tracking: An experimental survey. *IEEE Transactions on Pattern Analysis and Machine Intelligence* 36 (7):1442–1468.
van de Weijer, J., Schmid, C., Verbeek, J., & Larlus, D. 2009. Learning color names for real-world applications. *IEEE Transactions on Image Processing* 18(7):1512–1523.
Wibowo, S.A., Lee, H., Kim, E.K., & Kim, S. 2018. Collaborative learning based on convolutional features and correlation filter for visual tracking. *International Journal of Control Automation & Systems* 16(1):335–349.
Wu, Y., Lim, J., & Yang, M.H. 2015. Object tracking benchmark. *IEEE Transactions on Pattern Analysis and Machine Intelligence*, 37(9):1834–1848.

Calculation method of infrared temperature on the natural ground surface

Chen Shan & Ma Jun-chun
Xi'an Research Institute of High Technology, Xi'an, Shanxi, China

ABSTRACT: Infrared detection is one of the most important means of modern reconnaissance. The recognition method based on infrared image is widely used in infrared detection. In the process of infrared scene generation, infrared image generation of the natural ground surface is the key link. Starting with various factors affecting the boundary conditions of surface temperature, this paper firstly calculates the change of surface temperature by establishing the transient heat balance equation of the ground surface; secondly, combining with the principle of infrared imaging detection, the method of simulating and generating surface infrared image is given. By comparing the calculated value with the actual measured value, it is shown that the surface temperature curve calculated by the method proposed in this paper is quite accurate, and the infrared image generation method given is more reasonable.

1 INTRODUCTION

Infrared images of ground backgrounds are extremely widely used in remote sensing and military research. The source of traditional infrared images mainly relies on thermal imaging devices to conduct field tests for specific geographical backgrounds in specific environments. Although this method can obtain reliable infrared image data, the following problems exist: 1) expensive experimental costs; 2) due to geography, weather and other reasons, image data is relatively inadequate, which cannot meet the needs of users. Therefore, it is necessary to use some algorithms to obtain infrared image of ground background, which not only can avoid the above problems, and also can obtain thermal infrared simulation images of all-weather and various background areas.

Shuai-yang, ZHAO. 2018. Li-xia LIU. 2018. It is known from the basic law of infrared radiation that the infrared radiation of an object is determined by the surface temperature and the emissivity. Generally, once the target surface material is confirmed, the emissivity is determined. However, the surface temperature of the object is affected by various complicated factors. So how to determine the surface temperature is the key to simulate the infrared image of the ground surface.

In this paper, various environmental factors affecting the surface temperature are analyzed, and the boundary conditions for calculating the surface temperature are determined. The surface temperature varying with time is obtained by establishing the surface infrared simulation model, and compared with the measured results. On this basis, the surface infrared radiation is calculated, and the generation method of infrared simulation sequence images at different times of the day is analyzed.

2 INFRARED TEMPERATURE CALCULATION OF GROUND NATURAL SURFACE

2.1 *Heat exchange between the ground surface and the external environment*

The heat exchange between the ground surface and the external environment is usually carried out in three forms: radiation, convection and conduction. Because the conduction is mainly carried out between the ground surface and its interior, and the surface which is meaningful for calculating infrared radiation is mostly exposed in the air, the heat exchange between the ground surface and the external environment is mainly carried out by radiation and convection. The heat exchange between the ground surface and the external environment mainly includes solar short wave radiation, atmospheric long wave radiation, heat convection between the lower atmosphere and the ground surface, heat conduction from the ground surface to the underground, latent heat effect and radiation from the ground surface itself to the outside world.

2.1.1 *The solar short wave radiation*
The effects of solar radiation are mainly during the day, and their radiant flux varies with season, time, weather and geographical conditions. For the surface, solar radiation can generally be divided into two parts: direct solar radiation and scattering solar radiation.

Direct solar radiation refers to solar radiation that is directly received without changing direction, which can be given by the following formula:

$$I_{p,b} = rI_{sc}p^{m}\cos\theta_{T} \left[W/m^{2}\right] \qquad (1)$$

In the formula, r is the corrected value caused by the distance between the sun and the earth, I_{sc} is the solar constant, p is the atmospheric transparency, m is the air mass, and θ_{T} is the solar angle of incidence.

Scattering solar radiation refers to the received solar radiation which is reflected and scattered by the atmosphere and changes its direction. It can be derived from the following formula:

$$I_{p,d} = C_{1}(\sin h)^{C_{2}} \quad [cal/cm^{2} \cdot min] \qquad (2)$$

In the formula, C_{1} and C_{2} are empirical coefficients whose values depend on atmospheric transparency; h is the solar elevation angle.

Taking the factors of cloud occlusion into account, the solar radiation received by the surface is:

$$Q_{sun} = \alpha_{sun} \cdot CCF \cdot (I_{p,b} + I_{p,d}) \qquad (3)$$

In the formula: sun is the absorptivity of solar radiation on the surface; CCF is the cloud cover coefficient, which refers to the ratio of the sky area not occupied by clouds to the area of the whole sky area; for sunny skies, CCF = 1; for cloudy days, CCF = 0, and in most cases, the value of CCF is between 1 and 0.

2.1.2 *Atmospheric long wave radiation*
The radiation of the sky atmosphere is also a factor affecting the target temperature. After absorbing a certain amount of solar heat and the heat of the earth, the atmosphere has a certain temperature and thus radiates to the surface. The atmospheric radiation of the sky can also be equivalent to an infinitely large horizontal gray body plane located above, and the atmospheric radiation received by the ground surface is:

$$Q_{sky} = \alpha_{sky} \cdot C.C \cdot \sigma T_{sky}^{4} \left[W/m^{2}\right] \qquad (4)$$

In the formula: α_{sky} is the absorptive of surface to sky atmospheric radiation; C.C is the coefficient of cloud thickness; σ is the Boltzmann constant; T_{sky} is the atmospheric temperature.

2.1.3 *The self-radiation of the ground surface*

The heat loss from surface radiation to outer space can be obtained by Stefan-Boltzmann's law:

$$Q_{rado} = \varepsilon \sigma T^4 \tag{5}$$

In the formula: ε is the surface emissivity; T is surface temperature.

2.1.4 *Heat convection*

The heat exchange between the ground surface and air due to relative flow is as follows:

$$Q_{conv} = H(T_{air} - T) \tag{6}$$

In the formula: H is the heat convection coefficient; T_{air} is the low-altitude atmospheric temperature.

2.2 *The transient heat balance equation of the ground surface*

Because the vertical gradient of known temperature has the greatest influence on the ground surface temperature field, while the horizontal heat flux is relatively small and can be neglected approximately, the one-dimensional transient heat conduction mathematical model of the ground surface can be used in calculating the ground surface downward heat conduction. Chun-lei, WANG. 2014. When considering the upper boundary conditions of the ground surface, the complex heat exchange between the ground surface and the surrounding environment can be fully considered, that is to say, the heat balance equation of the ground surface.

Assuming that the heat exchange occurring on the ground surface and its interior is perpendicular to the ground surface, the heat balance equation of the ground surface can be established as follows:

$$\rho c \frac{\partial T}{\partial \tau} = \frac{\partial}{\partial z}\left(k \frac{\partial T}{\partial z}\right) \tag{7}$$

In formula: ρ is density; c is specific heat capacity; T is temperature; τ is time; K is thermal conductivity; z is depth coordinate.

2.3 *Initial-boundary value condition*

Initial value problem refers to the temperature at the initial time of the model area. Assuming that the weather conditions are constant every day, the temperature at 0 o'clock in a day should be equal to that at 24 o'clock. Therefore, the initial value condition can be set as follows:

$$T|_{\tau=0h} = T|_{\tau=24h} \tag{8}$$

The boundary condition refers to the connection or interaction of heat exchange between the heat conduction object on its boundary surface and the external environment. For unsteady heat conduction, it is often the external driving force for the process to occur and develop. Near the surface, the temperature varies greatly. The main influencing factors are various complex heat exchanges. When the ground surface reaches a certain depth, the temperature changes very slowly, which can be considered to have reached constant temperature. Therefore, the boundary conditions can be set as follows:

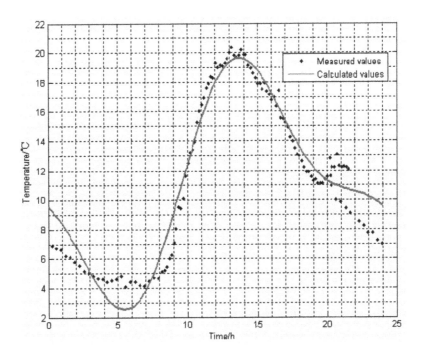

Figure 1. The measured values and calculated values of ground surface temperature.

Lower boundary $\qquad T|_{GroundBottom} = Const$

Upper boundary $\qquad k\dfrac{\partial T}{\partial z}\bigg|_{GroundSurface} = Q_{sun} + Q_{sky} + Q_{conv} - Q_{LE} - Q_{rado}$ $\qquad$ (9)

2.4 Calculation results and discussion

By using the finite difference method, the numerical form of the model is established from formula (7) to (9), and the numerical solution of the model is obtained. Taking an area in Xi'an as an example, as shown in Figure 1, the time of calculation and measurement is October 17, 2007. The weather is clear, the highest temperature is 21°C, the lowest temperature is 10°C, the relative humidity is 30%, the wind force is grade 2, and the ground surface material is sandy loam wet gray ground.

It can be seen from Figure 1 that the theoretical calculated value is consistent with the measured value at the temperature, and the temperature difference is relatively large at the initial time, which may be caused by artificially setting the iterative initial value during the calculation.

3 INFRARED IMAGE GENERATION OF NATURAL GROUND SURFACE

3.1 Calculation of surface infrared radiation

After calculating the ground surface temperature, the infrared radiation of the ground surface can be calculated. The infrared radiation of the ground surface mainly consists of two parts: the infrared radiation emitted by the ground surface itself and the infrared radiation reflected by the ground surface.

Considering the spectral characteristics of the target, according to Planck's law, the radiation of the ground surface in 1-2 bands is:

$$E_{\text{self}} = \varepsilon \int_{\lambda_1}^{\lambda_2} \frac{C_1}{\lambda^5 (e^{C_2/\lambda T} - 1)} \, d\lambda \tag{10}$$

In the formula: ε is the emissivity of ground surface; c_1 and c_2 respectively are the first and second radiation constants.

3.2 Generation of infrared image

Xia-lang MAO, 2014. Infrared image simulation is to simulate the thermal imaging system to simulate the infrared image. In the infrared scene, the radiation of the object detected by the infrared imaging detecting instrument in the scene, in addition to the radiation emitted by the object, there is also the environmental infrared radiation at the detection point. In addition, the factors of atmospheric attenuation should also be taken into account.

$$E_{\text{detector}} = \bar{\tau} \cdot (E_{\text{self}} + E_{\text{envi}} + E_{\text{reflect}}) \tag{11}$$

The infrared environmental radiation can be approximately replaced by the infrared sky long wave radiation received by the detection point. $\bar{\tau}$ is the average atmospheric transmittance, which can be calculated by the corresponding atmospheric transmission software.

4 CONCLUSION

In view of the characteristics of natural ground surface, this paper analyses various factors affecting the infrared radiation of natural ground surface, and focuses on the calculation of natural ground surface temperature. By establishing a simulation model, the calculated ground surface temperature is basically close to the measured temperature, which also has got good simulation results. Finally, a method for simulating infrared images is given. The next step is to get an infrared simulation image based on the background and target.

REFERENCES

Chun-lei, Wang. 2014. Algorithm Research about Retrieval of Land Surface Temperature with Estimated Emissivity. Journal of Hebei United University Natural Science Edition 20(3): 11–15.
Li-xia Liu. 2018. Simulation of global mid-infrared background based on remote sensing data. Infrared and Laser Engineering 47(11): 45–48.
Shuai-yang, Zhao. 2018. Analyses of Land Surface Emissivity Characteristics in Mid-Infrared Bands. Spectroscopy and Spectral Analysis 38(5): 20–24.
Xia-lang Mao. 2014. Retrieval land surface temperature from visible infrared imager radiometer suite data. Transactions of the Chinese Society of Agricultural Engineering 40(8): 24–27.

Advances in Optoelectronic Technology and Industry Development – Jose & Ferreira (Eds)
© 2020 Taylor & Francis Group, London, ISBN 978-0-367-24634-1

Research on gesture-recognition method in video based on the sparse representation theory

Yang Lei, Lu Feng, Lv Zhenglong & Wu Shiliang
School of Mechatronic Engineering and Automation, Shanghai University, China

ABSTRACT: Gesture recognition is an important research topic in computer vision. Existing gesture recognition methods are generally based on single image and lack spatiotemporal continuity in the analysis of image content. In order to deal with this problem, a new gesture recognition method in video based on sparse representation theory is proposed in this paper. Firstly, the foreground image of the hand region is obtained by using the skin color segmentation of the YCbCr color space for a continuous video. Secondly, the center of gravity of the foreground image for the hand region is extracted as feature vector for recognition. Gesture dictionary is further constructed, and a sparse representation model of certain kind of gesture is established. Then, gestures in video are classified by determining the sparse representation error for a new sample to be identified. Finally, experiments on the collected video sequences are performed. Experimental results show that the proposed method can recognize four kinds of gestures such as moving up, down, left and right in video. The proposed method would be used to recognizing more complex gestures in future work.

Keywords: gesture recognition, sparse representation, image segmentation, dictionary learning

1 INTRODUCTION

The research on gesture recognition began in the 1980s, and it is a typical multidisciplinary cross-knowledge study. With the development of computer hardware technology and the continuous improvement of artificial intelligence technology, gestures as the most direct and convenient means of human-computer interaction have been widely concerned by researchers. Gesture recognition research is mainly divided into data glove-based gesture recognition and computer vision-based gesture recognition (Pisharady P K & Saerbeck M 2015). The former requires wearable equipment and is therefore inconvenient and costly (Chen et al. 2018), so most of the gesture recognition methods are based on the latter. Gesture recognition based on computer vision first uses a common camera to capture gesture images or video, and then uses a computer recognition algorithm for gestures to identify dynamic gesture information (Yan 2018). In the early 1990s, many scholars began to study computer vision. Park et al. (2004) proposed a Hidden Markov Model (HMM) based gesture recognition method to enhance the difference between different categories and reduce the computational cost. Wilson proposed a state-based gesture recognition method, which divides gestures into a spatial motion trajectory into different states, and uses the results as a basis for recognition (Bobick A F & Wilson A D 1998). Zou et al. (2012) used motion detection and skin segmentation methods to extract gesture regions, then used Mean Shift tracking algorithm to obtain gesture trajectories, and used hidden Markov model to recognize gestures. Wang et al. (2014) extracted and smoothed the gesture contour through binarization and HDC processing, and found the gesture contour feature to complete the gesture image recognition. The method is simple, but the robustness is poor. Yang et al. (2014) proposed an ultrasonic-based gesture recognition method, which can realize the recognition of 24 gestures on the mobile intelligent platform, and the average recognition rate is 93%. However, the accuracy of the recognition accuracy in the complex noise environment is different. Wan et al. (2013, 2014) extended the scale-invariant feature transform (SIFT) feature to obtain 3D enhanced motion SIFT (EMo SIFT) and 3D sparse action SIFT (SMo SIFT), and through sparse keypoint blending features (MFSK) for gesture recognition (Cheng 2009). Cao et al. (2015) first segmented the gesture image in the HSV color space, and then realized the gesture image recognition based on the extracted gesture invariant features. The recognition accuracy of this method is high, but the recognition cost is large. Tao et al. (2016) firstly performs gestures such as YUA skin segmentation and connectivity detection to obtain complete gesture features, and then uses the support vector machine to construct a classifier that classifies different gestures to realize gesture image recognition. The efficiency of recognition of this method is high, but the recognition accuracy is not high. Weng and Zhan (2012) used color features to detect skin color regions to segment human hands, detect fingertips based on human hand contours and convex defects, and use gestures to indicate gestures, thus completing gesture recognition.

In recent years, researchers gradually shifted their research focus to gesture recognition in video, which contains spatiotemporal continuity in the analysis of image content. At the same time, sparse representation theory can better highlight the essential characteristics of signals than traditional recognition methods. It has been widely used in image processing. In this paper, the sparse representation is introduced into the gesture recognition of computer vision. A new gesture recognition in video based on sparse representation is proposed.

2 PREPROCESSING OF VIEDO

The gesture recognition process based on computer vision is roughly divided into four steps, as shown in Figure 1.

Before the gesture recognition, the acquired gesture image is pre-processed, including two main steps of gesture segmentation and feature extraction. In this paper, the threshold adaptive method based on YCbCr color space is used to segment the gesture and the first-order moment estimation algorithm is used to extract the center of gravity of the gesture.

2.1 Hand detection and segmentation based on YCbCr color space

YCbCr color space model is widely used in image and video processing, machine vision and other fields.The conversion formula of the RGB color space model linearly transforming to form the YCbCr color space model is as follows:

$$\begin{cases} Y = 0.299R + 0.587G + 0.114B \\ Cb = -0.1687R - 0.3313G + 0.500B + 128 \\ Cr = 0.500R - 0.4187G - 0.0813B + 128 \end{cases} \tag{1}$$

There are two advantages to using the YCbCr color space. First, the linear conversion from the RGB space of the original image to the YCbCr space is relatively easy to implement and the calculation speed is faster. Second, the YCbCr space is a color space in which the luminance is separated, and the chrominance signals Cb, Cr and the luminance signal Y are independent of each other. The color distribution of the skin color in the Cb-Cr space is better. When the prior skin color model is established, only the Cb and Cr components are taken, and the interference of the light condition on the skin color segmentation can be reduced in a certain procedure.

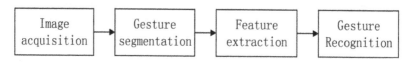

Figure 1. Basic process framework for gesture recognition.

The Cb and Cr channels are thresholded in the YCbCr space to define the gesture skin area. According to the literature (Hsu R L et al. 2002), the chromaticity decision range suitable for Asian skin color is Cr[133,173] and Cb[77,127].

2.2 *Feature detection of the gesture based on the center of gravity of hand*

The color and texture of different gestures are basically the same for the same user. However, the traits and contours vary with the gesture.The features that can be used for gesture recognition are only shapes and contours. Hu proposed the concept of moment invariant, and the nonlinear combination of moments can obtain moment invariants with translation invariance, rotation invariance and proportional invariance (Hu M K 1962). Since the gesture samples have changes in translation, size, and rotation during the acquisition process, the Humoment feature can be used to recognize the gesture. For a gesture image $f(x,y)$, its $(p+q)$ order two-dimensional origin moment is defined as

$$M_{pq} = \int_{-\infty}^{+\infty}\int_{-\infty}^{+\infty} x^p y^q f(x,y)dxdy \quad p,q = 0,1,2\dots \tag{2}$$

The corresponding center moment is

$$\mu_{pq} = \int_{-\infty}^{+\infty}\int_{-\infty}^{+\infty} (x-\overline{x})^p (y-\overline{y})^q f(x,y)dxdy \tag{3}$$

The purpose of this paper is to identify the dynamic gestures in the video. In the gesture segmentation stage, the complete gesture area is extracted, and then the center of gravity of the image is found. Since the center of gravity of the gesture moves with the gesture, the motion track of the center of gravity of the gesture is the motion track of the gesture.

Gesture motion feature extraction based on gesture center of gravity tracking is mainly divided into the following steps: 1) Extract the key points of the center of gravity of the gesture according to the time series, extract the center of gravity of each frame, and calculate the coordinate position of the center of gravity of each frame image in the entire image (D. Gong et al. 2013); 2) Processing of barycentric coordinate data, including normalization and elimination of bad feature points; 3) Gesture trajectory extraction, the trajectory fitting of the extracted center of gravity feature points is obtained to obtain the center of gravity motion curve, that is, the gesture motion trajectory.

3 GESTURE RECOGNITION BASED ON THE SPARSE REPRESENTATION THEORY

3.1 *The sparse representation theory*

Sparse representation is a research direction in the field of signal processing since the 1990s (R. Vidal 2009). Since it has achieved good application results in image processing and speech processing, it has received extensive attention (E. Elhamifar & R. Vidal 2009).

Suppose there is an observation signal $y \in R^{m\times1}$, an overcomplete dictionary $D = [d_1, d_2, \cdots d_j, \cdots d_n] \in R^{m\times n}$, $(d_j \in R^{m\times1}$, $j = 1\cdots n$, $m \ll n)$, then the signal y is linearly represented by D such as:

$$y = Dx = x_1 d_1 + x_2 d_2 + \dots + x_n d_n \tag{4}$$

where $x \in R^{n \times 1}$ is the coefficient vector in D for linear representation y. It can be seen from $m < n$ that Equation 4 is an underdetermined equation, and the same solution is difficult. If x is sparse, i.e., the number of non-zero elements in x is as small as possible, the coefficient vector x can be obtained by solving following equation:

$$\min_{x \in R^n} \|x\|_0 \quad s.t \quad y = Dx \tag{5}$$

where $\|\cdot\|_0$ is the l_0-norm and is used to measure the number of non-zero elements in the vector. The l_0-norm sparse representation model is NP-hard, and since the l_1-norm is convex, the l_1-norm can be used instead of the l_0-norm approximation (G. Liu et al. 2010). Replace all the models represented by the l_0-norm above with the l_1-norm representation:

$$\min_{x \in R^n} \|x\|_1 \quad s.t \quad y = Dx,$$

$$\min_{x \in R^n} \|x\|_1 \quad s.t \quad \|y = Dx\|_2$$

and

$$\hat{x} = \arg\min_{x \in R^n} \|y - Dx\|_2^2 + \lambda \|x\|_1 \tag{6}$$

3.2 Solution of the sparse representation model

In this paper, the Alternating Direction Method of Multipliers (ADMM) (E. Elhamifar & R. Vidal 2013) is used to solve Equation 6. First, Equation 6 is written as ADMM:

$$\min \|x\|_1 \\ s.t. \ y - Dx = 0 \tag{7}$$

By applying the augmented Lagrangian multiplier to eliminate the equality y-Dx constraint in the above equation, so an augmented Lagrangian expression is written as:

$$L(x, Y, \mu) = \|x\|_1 + \langle Y, y - Dx \rangle + \frac{\mu}{2} \|y - Dx\|_2^2 \tag{8}$$

where $\mu > 0$ is an over-regularization parameter and is Y a Lagrangian multiplier. The model uses ADMM to solve the above Lagrangian function, which is iterated by the following four steps (where j represents the number of iterations):

1) Solve x

$$x^{j+1} = \arg\min_{x} \|x\|_1 + <Y^j, y - Dx> + \frac{\mu^j}{2} \|y - Dx\|_2^2 \\ = (D^T D)^{-1} \left(D^T y + \frac{D^T Y}{\mu} - \frac{I}{\mu} \right) \tag{9}$$

2) Solve Y

$$Y^{j+1} = Y^j + \mu(y - Dx^{j+1}) \tag{10}$$

28

3) Solve μ

$$\mu^{j+1} = \max(\mu^j \rho, \mu_{\max}) \tag{11}$$

where $\rho > 1$ is a constant.

The ADMM is used to iteratively solve the above three variables until $\|y - Dx\|_2$ is less than the specified error, and the sparse representation coefficient x can be solved.

3.3 *Gesture classification using sparse subspace clustering*

From a geometric point of view, many of the pattern classes to be identified can be represented by specific subspaces, each of which represents a category (P. Doll'ar et al. 2005). The different gesture motion trajectories are considered to be distributed in different subspaces (A. Sanin et al. 2013), and the recognition of the gesture trajectory can be transformed into a sparse representation classification model to solve (X. Zhang et al. 2014).

Suppose that the training sample consisting of a certain type of gesture is $A_i = \left[a_{i1}, a_{i2}, \mathrm{k} a_{ij}, \mathrm{k}, a_{in_i}\right] \in R^{m \times n_i}$, where $m = w \times h$, a_{ij} is a column vector, $j = 1, 2, \ldots, n_i$, which is used to represent a certain type of gesture image of size $w \times h$ in the training sample. If n_i is large enough, then the new gesture sample $y_i \in R^{m \times 1}$ belonging to the i-th class can be approximated by the linear combination approximation of the i-th training sample:

$$y_i \approx A_i x_i \tag{12}$$

where $x_i \in R^{n_i \times 1}$ represents the sparse representation coefficient on Ai. When the training sample $A_i = [A_1, A_2, \cdots, A_c] \in R^{m \times n}$ (n indicates all training samples) contains c categories, any type of test sample y can be sparsely represented on the entire training sample as:

$$y = Ax \tag{13}$$

where x is a sparse vector which is very sparse. In theory, only the coefficient corresponding to the training sample of y is not 0 in x, and the remaining coefficients in x are all 0. When $m > n$, the system of equations is considered to be overdetermined, i.e. the vector x is uniquely determined. The x in the l_0-norm satisfies the minimum value of $\|x\|_0$, which is expressed by the following formula:

$$\hat{x} = \arg\min \|x\|_0 \; s.t. \; y = Ax \tag{14}$$

The above formula is still an NP-hard problem. It is indicated by sparse representation and compressed sensing theory that the following convex optimization approximation can be used:

$$\hat{x} = \arg\min \|x\|_1 \; s.t. \; y = Ax \tag{15}$$

The $\|x\|_1$ in the above equation is x of the l_1-norm, and the problem can be expressed as a tolerance error form by standard linear equation optimization:

$$\hat{x} = \arg\min \|x\|_1 \; s.t. \; \|y - Ax\|_2 \leq \varepsilon \tag{16}$$

where ε is the tolerance error, the test sample y is classified into the category with the largest coefficient, that is, the highest correlation, and the category to which the test sample belongs is determined according to the residual minimum value classification, namely:

$$identity(y) = \arg\min \|y - Ax\|_2 \tag{17}$$

4 EXPERIMENTAL RESULTS

4.1 *Design of experiments*

First, the gesture videos of the front, back, left, and right directions are recorded, which contain 50 samples in each class, each sample duration is 8 seconds, and 40 videos of each class are used to construct training samples, and each training video is extracted. As for one-dimensional feature vectors, the training samples extracted from the gestures in the four directions can form 4 categories, and the overcomplete dictionary is 200 columns. This experiment was implemented using the 2016 version of **MATLAB** software. The main steps of the experimental design are as follows:

1) Split each video into 200 frames, convert each frame of **RGB** color space into YCbCr color space, take the Cb and Cr components of YCbCr color space, and segment the foreground by the optimal threshold of skin color. Gestures and convert foreground gestures into binary images. Finally, the binary image is processed by morphology to obtain a foreground image with better effect.
2) Calculate the coordinates of the center of gravity of the foreground image gesture in each frame by the first moment. Each video sequence will have 200 coordinates of the center of gravity coordinates, and then the coordinates of the center of gravity will be drawn in a time axis diagram to extract the feature vector.
3) The trained gesture center of gravity feature is constructed into a complete dictionary, and the gesture direction is determined by calculating the test gesture sparse subspace representation and the reconstruction error.

4.2 *Results of foreground extraction and trajectory of the center of gravity*

Figures 2 and 3 are a group of upward and leftward video gesture sequences and their foreground extraction images, respectively.

After extracting the foreground gesture result map, according to the method proposed in this paper, the center of gravity of the foreground image is obtained, and the center of gravity trajectory of the entire 8-second video is drawn. The trajectory of the center of gravity in the upper and lower directions is shown in Figure 4 (the upper is the red curve, the lower is the green curve), and the reflection of the y-axis coordinates is reflected in the figure. The comparison of the center of gravity trajectories in the left and right directions is shown in Figure 5 (left is the blue curve, right is the black curve), and the reflection of the x-axis coordinates is reflected in the figure.

4.3 *Experimental results and analysis*

It can be seen from the trajectory of the center of gravity of the experimental results that the amplitude of the center of gravity trajectory in the up and down direction is significantly different from the trajectory of the center of gravity in the left and right direction. Based on this obvious distinguishing feature, we can extract the center of gravity feature vector by using the magnitude of the amplitude variation, and then set the threshold of the corresponding amplitude change to distinguish the up and down direction from the left and right direction.

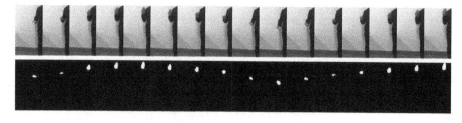

Figure 2.   Upwards gesture color image and foreground extraction gesture.

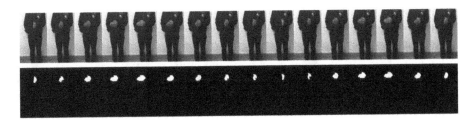

Figure 3.   Leftwards gesture color image and foreground extraction gesture.

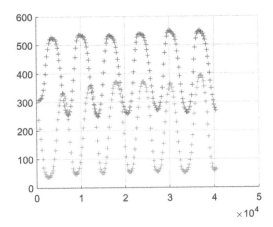

Figure 4.   Up and down center of gravity trajectory.

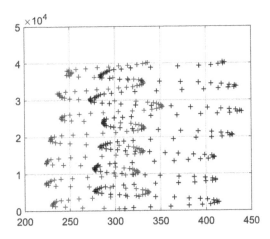

Figure 5.   Left and right center of gravity trajectory.

In this experiment, the difficulty in judging the direction of the gesture is to distinguish the direction of motion of the up and down center of gravity trajectories, and the direction of motion of the left and right center of gravity trajectories. Through the analysis of sample data in this experiment, we find that the gravity trajectories present a relatively dense state at peaks and troughs, respectively. Therefore, we use the frame difference method for each set of 400*1 training samples, that is, each line of the new sample consists of the inter-frame difference of the original samples. The new sample thus obtained can reflect the trend of gesture movement. Taking the upward gesture sample as an example, the difference between frames is large during the upward movement of the hand, and the difference between the frames tends to become smaller and more densely distributed when the hand reaches the top. The new sample

31

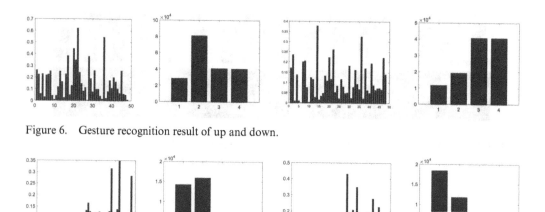

Figure 6.   Gesture recognition result of up and down.

Figure 7.   Gesture recognition result of the left and right.

is used as a dictionary atom to construct a gesture-sparse overcomplete dictionary. The next step is to use a number of samples as tests to classify and identify gestures. The recognition results of one of the groups are shown in Figures 6 and 7. The left pictures show the sparse coefficient distribution, and the right pictures shows the sparse reconstruction error. 1, 2, 3, and 4 are the reconstruction errors of the up, down, left, and right, respectively. According to the size of the reconstruction error, the smaller decision is the direction of the gesture.

In the experiment, the recognition of the up, left and right gestures is successfully completed, and the down gesture identifies the failure. It is not difficult to see that although the recognition fails, the reconstruction errors of the up and down are relatively close. A total of 10 sets of gestures (up, down, left and right) were selected as test samples for gesture recognition. The recognition rates were 80% on the up, 70% on the down, 80% on the left, and 70% on the right.

The analysis of the experimental results shows that the proposed method can distinguish between up and down and left and right gestures, but as shown in the experimental results, some problems are encountered in distinguishing between up and down, and left and right (downward gesture recognition fails, the left and right reconstruction errors are not significantly different). In the following, we will carry out further research on feature extraction. We consider extracting dense regions of gravity center coordinates, discarding discrete points, such as selecting 0–20% and 80%–100% thresholds to segment the gravity center trajectory, and obtain each peak. Several barycentric coordinate regions centered on the trough. The variance of the barycentric coordinate set in each region is calculated separately. Theoretically, the upward gesture center of gravity trajectory map is taken as an example, and the variance of the barycentric coordinate set at the trough should be smaller than the variance at the peak. The new one-dimensional feature vector is formed by the above variance, and is used as an atom of the overcomplete dictionary to form a new gesture dictionary, so that the up, down, left and right gestures should be better recognized. This will be a research direction in our future work.

5   CONCLUSION

In this paper, four kinds of gesture in video named up, down, left and right of hand is proposed, which is based on calculating the trajectory of the gravity center of a hand part of a video by first moment function. Firstly, the data set of the four motion directions of the gesture is achieved, and the hand part in the video is extracted by color space transformation. Secondly, the first moment is used to calculate the center of gravity of each gesture, and a series of gesture center of gravity coordinates are calculated in chronological order for comparison. After comparing the experimental data, it can be clearly seen that there is a significant difference in amplitude variation between the up and down direction and the left and right direction. The results can be given by setting the corresponding thresholds. However, the distinction between the up and down directions and the left and right directions is not obvious. It will be the future direction for us to distinguish complex gesture movements of hands.

REFERENCES

A. Sanin, C. Sanderson, M.T. Harandi, and B.C. Lovell, "Spatiotemporal covariance descriptors for action and gesture recognition," 2013 IEEE Workshop on Applications of Computer Vision (WACV), pp. 103–110, IEEE, 2013.

Bobick A.F, Wilson A.D. A State-Based Approach to the Representation and Recognition of Gesture [J]. IEEE Transactions on Pattern Analysis and Machine Intelligence, 1998, 19(12): 1325–1337.

Cao Xiang, Chen Xiang, Su Ruiliang. Optimized strategy of fHMM for real-time multi-sensor gesture recognition [J]. Space Medicine & Medical Engineering, 2015(3): 183–189.

Chen Guoliang, Ge Kaikai, Li Conghao. Complex dynamic gesture recognition based on multiple features and HMM fusion [J]. Journal of Huazhong University of Science and Technology (Natural Science), 2018(12).

Cheng Wenshan. Research of gesture recognition based on skin color and camshift algorithm [D]. Huazhong Normal University, 2009.

D. Gong, G. Medioni, and X. Zhao. Structured time series analysis for human action segmentation and recognition. IEEE Trans. Pattern Anal. Mach. Intell., vol. 36, no. 7, pp. 1414–1427, Jul. 2013.

E. Elhamifar and R. Vidal, "Sparse subspace clustering," in Proc. IEEE Conf. Comput. Vis. Pattern Recognit., Jun. 2009, pp. 2790–2797.

E. Elhamifar and R. Vidal. Sparse subspace clustering: Algorithm, theory, and applications. IEEE Trans. Pattern Anal. Mach. Intell., vol. 35, no. 11, pp. 2765–2781, Nov. 2013.

G. Liu, Z. Lin, and Y. Yu, "Robust subspace segmentation by low-rank representation," in Proc. Int. Conf. Mach. Learn., 2010, pp. 663–670.

Hsu R.L, Abdel-Mottaleb M, Jain A.K. Face Detection in Color Images [J]. Journal of Chengdu Textile College, 2002, 24(5): 696–706.

Hu M.K. Visual Pattern Recognition by Moment Invariants [J]. Information Theory, IRE Transactions on, 1962, 8(2): 179–187.

P. Doll'ar, V. Rabaud, G. Cottrell, and S. Belongie, "Behavior recognition via sparse spatio-temporal features," 2nd Joint IEEE International Workshop on Visual Surveillance and Performance Evaluation of Tracking and Surveillance, pp. 65–72, IEEE, 2005.

Park H.S, Kim E.Y, Jang S.S, et al. An HMM Based Gesture Recognition for Perceptual User Interface [C] Pacific Rim Conference on Advances in Multimedia Information Processing. Springer-Verlag, 2004.

Pisharady P.K, Saerbeck M. Recent methods and databases in vision-based hand gesture recognition: A review [J]. Computer Vision & Image Understanding, 2015, 141(C): 152–165.

Qifan Y, Hao T, Xuebing Z, et al. Dolphin: Ultrasonic-Based Gesture Recognition on Smartphone Platform [C] 2014 IEEE 17th International Conference on Computational Science and Engineering (CSE). IEEE Computer Society, 2014.

R. Vidal, "Subspace clustering," IEEE Signal Process. Mag., vol. 28, no. 2, pp. 52–68, Mar. 2011.

Tao Meiping, Ma Li, Huang Wenjing, et al. A gesture recognition research based on unsupervised feature learning [J]. Microelectronics & Computer, 2016, 33(1): 100–103.

Wan J, Ruan Q, Li W, et al. One-shot Learning Gesture Recognition from RGB-D Data Using Bag of Features [J]. Journal of Machine Learning Research, 2013, 14(1):2549–2582.

Wan J, Ruan Q, Li W, et al. 3D SMoSIFT: Three-dimensional sparse motion scale invariant feature transform for activity recognition from RGB-D videos [J]. Journal of Electronic Imaging, 2014, 23(2).

Wang Zhenshui, Li Lin, Liu Xiaoping. Research on multi-granularity dynamic gesture recognition method of supporting self-definition [J]. Journal of Electronic Measurement and Instrument, 2014, 28(4): 416–423.

Weng Hanliang, Zhan Yinwei. Multi-feature gesture recognition based on vision [J]. Computer Engineering and Science, 2012, 34(2): 123–127.

X. Zhang, Y. Yang, H. Jia, H. Zhou, and L. Jiao, "Low-rank representation based action recognition," 2014 International Joint Conference on Neural Networks (IJCNN), pp.1812–1818, IEEE, 2014.

Yan Shiyang. Dynamic gesture recognition method based on computer vision [D]. Nanjing University of Posts and Telecommunications, 2018.

Zou Jiehua. Research on dynamic gesture trajectory recognition system based on monocular vision [D]. Xi'an: Xidian University, 2012.

Laser Technology and Applications

Advances in Optoelectronic Technology and Industry Development – Jose & Ferreira (eds)
© 2020 Taylor & Francis Group, London, ISBN 978-0-367-24634-1

Technical and analytical note on the performance maximization of spin lasers by optimizing the spin polarization

Ritu Walia & Kamal Nain Chopra*

Department of Physics, Maharaja Agrasen Institute of Technology, GGSIP University, Rohini, New Delhi, India

*Formerly Laser Science and Technology Centre (LASTEC), DRDO, Delhi, India, and

Photonics Group, Applied Optics Division, Department of Physics, Indian Institute of Technology, Hauz Khas, New Delhi, India

ABSTRACT: A technical analysis for the performance maximization of spin lasers by optimizing the spin polarization has been presented in this paper. Fine structure state for Mg atom and its decomposition into a superposition of various spin states of valence electrons, and control of the degree of spin polarization through time delay between laser pulses (ns and fs) have been described. Maximization of spin polarization, as a function of various quantities of dependence, has been suggested. The generation of thermoelectrically induced spin torque, and the mechanisms of spin torque generation have also been technically discussed.

Keywords: spin lasers, spin polarization, thermoelectrically induced spin torque, mechanisms of spin torque generation

1 INTRODUCTION

Spin lasers are based on the concept of spintronics, a new branch of electronics in which electron spin, in addition to charge, is manipulated to yield a desired electronic outcome. It is important to realize that all the spintronic devices act according to the simple scheme: (i) information is stored (written) into spins as a particular spin orientation (up or down), (ii) the spins, being attached to mobile electrons, carry the information along a wire, and (iii) the information is read at a terminal. Spin orientation of conduction electrons survives for a relatively long time (nanoseconds, compared to tens of femtoseconds during which electron momentum and energy decay), which makes spintronic devices particularly attractive for memory storage and magnetic sensors applications, and, potentially for quantum computing, where electron spin would represent a bit (called qubit) of information.

The field of spintronics is based upon the use of the direction of spin of electrons rather than their charge (positive or negative). Spin-polarized electrons can be considered to have two states; spin-up or spin-down, which can be used to represent on and off state respectively. Chopra (2017) has given a detailed analytical treatment of the theory and design aspects for the modeling and optimization of the efficiency of spin lasers.

A spin wave laser (Figure 1) is based on the emission of energy in the form of electromagnetic waves by electrons with axial and orbital spin undergoing transition from the higher energy spin states to the lower energy spin state. For a spin wave laser with a population inversion of spin states, the individual spins precess resulting in the stimulus to drop to a lower energy spin state, in the form of electromagnetic waves, matching the frequency of precession – the Larmor frequency. Interestingly, the spins

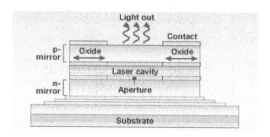

Figure 1. Spin laser showing the design of laser cavity.

are stimulated to emit electromagnetic waves, which are in phase with the stimulating electromagnetic waves.

2 MATHEMATICAL MODELING AND OPTIMIZATION OF SPIN POLARIZATION

Chopra (2014) has discussed the mathematical aspects of spin-related phenomena models and the associated criteria for studying the performance of spintronics devices in general, and spin lasers in particular. A good technique is based on recombination of holes and the spin relaxation within the active region of a p-doped semiconductor. Presuming a sufficient number of holes for recombination, for example, in a p-doped semiconductor, the spin relaxation within the active region can be accounted by using the well established approach (Dyakonov, 2008), in which the effectively measured degree of circular polarization is given by the following expression:

$$P_{circular,\ effective} = \left\{ \frac{P_{circular}}{1 + \left(\frac{\tau_{electron}}{\tau_S}\right)} \right\} \tag{1}$$

where τ_S is the spin lifetime, $\tau_{electron}$ is the electron lifetime, and $P_{circular}$ is the degree of circular polarization. The importance of this relationship can be judged from the fact that the impact of spin relaxation in the active medium is not determined by the spin lifetime alone, but by the electron-to-spin-lifetime ratio which has to be minimized, for the optimized results. Clearly, this can be done either by a long spin relaxation time or by a short electron lifetime. The spin laser designer has also to consider the fact that to ensure a minimized spin relaxation during transport, and thereby to maximize the performance of the system, all transport path lengths in spin-optoelectronic devices have to be kept as short as possible, and also within the permissible limits.

The polarization dynamics for the two heavy-hole(hh)-related circularly polarized transitions, considering two distinguished carrier densities for spin-up and spin-down carriers can be described by following the well established approach (Gahl et al., 1999), and the dynamic spin-flip model (SFM), which is based on a four-energy-level approximation, and takes only transitions between electron and heavy-hole states into account. Interestingly, these two carrier reservoirs are coupled by the spin relaxation rate, which in fact describes all kinds of microscopic spin relaxation processes by means of a single phenomenological parameter (Martin et al., 1997). Another important design consideration is that the circularly polarized light fields $E_\pm$ are coupled by the cavity anisotropies birefringence (γ_p) and dichroism (γ_a). The coupled rate equations are as given below:

$$\begin{aligned} E_\pm{}^{\cdot} = \{ &\kappa\,(1 + i\alpha)(N \pm m_Z - 1)E_\pm \\ &- (\gamma_a \pm i\gamma_p)E_\mp + \xi_\pm \sqrt{\beta\gamma\,(N \pm m_Z)} \} \end{aligned} \tag{2}$$

$$N^{\cdot} = \gamma[\eta_{+} + \eta_{-} - (1 + I_{+} + I_{-})N$$
$$- (I_{+} - I_{-})m_{\mp}] \tag{3}$$

and

$$m_{Z}^{\cdot} = \gamma(\eta_{+} - \eta_{-}) - [\gamma_{S} + \gamma(I_{+} + I_{-})]m_{Z}$$
$$- \gamma(I_{+} - I_{-})N_{+} \tag{4}$$

where $I_{\pm}$ is the intensity of the circularly polarized optical laser modes, which can be described by the complex amplitudes of the circularly polarized light fields by using the well known expression $I_{\pm} = |E_{x}|^{2}$, κ is the cavity decay rate related to the photon lifetime (Martin et al, 1997) by $(1/2\kappa)$, and α is the linewidth enhancement factor. The designer has also to take into account the influence of the spontaneous emission to the laser mode by using the spontaneous emission factor β, and the spontaneous emission noise terms $\xi_{\pm}$, which are described by complex Gaussian shaped distributions. Another important consideration is that the optical pumping and the electrical pumping can be modeled by using the pump terms $\eta_{\pm}$. Interestingly, the optical gain is implemented in the model by using a simple linear dependence of the population inversion. It has also to be noted that because of the optical gain for the circularly polarized light intensities being proportional to $(N \pm m_{Z} - 1)$, the gain values for I_{+} and I_{-} are unequal in case of a carrier spin polarization.

It has been theoretically established by now that the increased linewidths result from a coupling between intensity and phase noise, caused by a dependence of the refractive index on the carrier density in the semiconductor. The linewidth enhancement factor α is used to quantify this amplitude–phase coupling mechanism; and in fact is a proportionality factor relating phase changes to changes of the gain, which is given by the expression: $\Delta\varphi = \left(\frac{\alpha}{2}\right)\Delta g$. It has to be noted that the factor $\left(\frac{1}{2}\right)$ has been introduced to convert the change of power gain Δg to the change of amplitude gain $\Delta\varphi$. Interestingly, it is possible to calculate the factor α of a semiconductor for a given carrier density from a band structure model, which, however, is quite difficult, as the various parameters are interdependent. It is good to have an idea about its value, which for typical quantum wells, is of the order of $2 - 5$.

Spin polarization is defined as the degree to which the spin, i.e., the intrinsic angular momentum of elementary particles, is aligned in a given direction. This property is related to the spin, and therefore, to the magnetic moment of conduction electrons in ferromagnetic metals, like iron, resulting in spin-polarized currents. It is classified as static spin waves, or as preferential correlation of spin orientation with ordered lattices, which is the case in semiconductors or insulators.

Spin polarization of electrons is also produced by the application of a magnetic field. It is to be noted that the Curie law is used to produce an induction signal in Electron spin resonance (ESR). Spin polarization plays an important role in spintronics, a branch of electronics, and a lot of research work is going on the magnetic semiconductors as possible spintronic materials. (Ga,Mn)As at low temperatures is a commonly used material. Spin polarization of holes P in (Ga,Mn)As at low temperatures has been measured by the Andreev reflection on Ga/Ga0.95Mn0.05As junctions as high as ~80%. Similar effective P value of 77% has been reported from the magnitude of Tunneling Magneto Resistance (TMR) = 290% at low temperatures. It is interesting to note that these results agree well with the theoretical calculations of (Ga,Mn)As. Spin-polarization can also be usefully achieved by short laser pulses — two-electron system.

Spin-dependence of various quantities $(q_1, q_2, q_3, \ldots, q_n)$, f, provides more information on the dynamics

$$f(q_1, q_2, q_3, \ldots, q_n) = \frac{1}{2}\sum\nolimits_{m_s = \pm 1/2} f(q_1, q_2, - - -ms, - - -, q_n) \tag{5}$$

The designer has to maximize the total spin polarization function by differentiation as given below:

$$\frac{f_i\,(q_i)}{d\,q_i} > 0 \qquad (6)$$

The computations for the maximization of this function has to be done separately for all q_is. Nakajima (2004) has presented a generic pump-probe scheme to control spin polarization of photoelectrons/photo ions by short laser pulses. As reported, by coherently exciting fine structure manifolds of a multi-valence-electron system by the pump laser, a superposition of fine structure states is created. It has to be noted that (i) the spin-orbit effect is due to the electrostatic field of the electron and not the magnetic field created by its orbit, and (ii) the interaction between the magnetic field created by the, electron and the magnetic moment of the nucleus is a slight correction to the energy levels known as the hyperfine structure. It has been discussed that due to the fact that each fine structure state can be further decomposed into a superposition of various spin states of valence electrons, each spin component evolves differently in time, which implies that varying the time delay between the pump and probe lasers leads to the control of spin states. Some useful theoretical results have been presented for two-valence-electron atoms, particularly for Mg, which show that not only the degree of spin polarization, but also its sign can be manipulated through time delay. For Mg atom, the results for spin polarization dependence on time delay have been reproduced in Figure 2.

It can be seen that the spin polarization is maximum (equal to unity) at detuning factor $\Delta^{-1} = 7.5$ units (fs), and falls on both sides of this value, which is just expected as the maximum is always achieved at a particular value of detuning. Also, the fall is steeper for lower values, and touches zero at $\Delta^{-1} = 2.5$ units (fs). These values agree well with the theoretically predicted values. This implies that it is possible to control of spin states by varying the time delay between the pump and probe lasers. Specific theoretical results (Nakajima, 2004) have been reported for two-valence-electron atoms, in particular for Mg, which demonstrate that not only the degree of spin polarization, but also its sign can be manipulated through time delay. This agrees well with the findings of Bouchene et al. (2001), who have presented a theoretical analysis of the possibilities of producing spin-polarized electrons with a sequence of two ultra-short time-delayed laser pulses. For their study, the first pulse, which is right circularly polarized, is used to excite resonantly the fine structure np level of potassium atoms leading to a spin-flip, and thus polarizing the atom; and the second pulse is used to ionize the system, which results in the release of the spin-polarized electron. It is interesting to note that they have been able to examine and compare several situations corresponding to different polarizations of the ionizing pulse e.g. σ+, σ-and π; and also for each case, they have derived

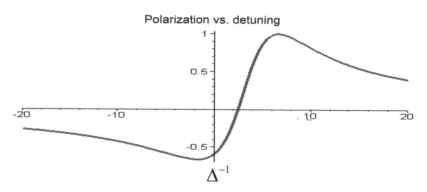

Figure 2. Degree of spin polarization vs detuning. Figure modified from the figure courtesy of Takashi (2004).

analytical expressions for the angular and global electron spin-polarization rates as well as for the differential cross sections. Their most important observation is that the obtained rates can be very high, even up to 100%, and can be controlled accurately on the femtosecond time scale by varying the time delay between the pulses. For the sake of comparison, these results have been reproduced in Figure 3.

The new feature observed in this case is that the degree of spin polarization can be varied between ~ 1 and 0 repeatedly in a similar fashion.

It has to be noted that spin-dependent effects are related to (i) spin-polarized electrons, connected with high energy physics, atomic and molecular processes, surface physics, and semiconductor physics; (ii) electron spin-polarized ions, connected with surface physics; and (iii) atomic and molecular processes, and nuclear-spin-polarized (doped) atom, connected with nuclear physics. It has also to be appreciated that (i) electron spin-polarization upon photoionization of rare gas atoms by $UV \sim VUV$ pulse; (ii) simultaneous production of spin-polarized electrons/ions with ns pulses; and (iii) ultrafast spin polarization; are also important in such studies.

It has been reported that coherent excitation of fine structure manifolds the spin-orbit coupling time (τ), and for the ultrafast pulses, spin-orbit coupling time is given by:

$$\text{spin-orbit coupling time} \sim \Delta^{-1} \qquad (7)$$

If pulse duration $\tau \ll \Delta^{-1}$, the system does not see spin-orbit interaction during the pump pulse. The generation of thermoelectrically induced spin torque is based on the spin polarization and spin currents, which is exerted on magnetization, and thus is used to control the state of a magnetic moment. Therefore, the mechanisms of spin torque generation are related to the (i) spin currents and (ii) spin polarization. The first type is based on the transfer of angular momentum from spin current to magnetization, which acts in systems with spatially non-uniform magnetization, and thus the absorbed spin current produces a spin-transfer torque, which in turn leads to many effects including switching orientation of magnetic moments, excitation of stationary precessions supported by current, and domain wall shifts. The second type of mechanisms resulting in the manipulation of magnetic moments is based on an effective magnetic field due to the spin–orbit interaction. This means that the electric field or temperature gradient in the presence of spin–orbit interaction induces a spin polarization, and in turn the induced spin polarization modifies the local magnetization, as discussed by Kurebayashi et al. (2014), who have explained that magnetization switching at the interface between ferromagnetic and paramagnetic metals, controlled by current-induced torques, has a good potential for use in magnetic memory technologies. It is important to understand the role played (i) in the switching by the spin Hall effect in the paramagnet, and (ii) by the spin–orbit torque originating from the broken inversion symmetry at the interface. The anti-damping components of these current-induced torques act against the equilibrium-restoring Gilbert damping of the magnetization dynamics. Kurebayashi et al. (2014) have reported the observation of an anti-damping spin–orbit torque that stems from the Berry curvature, in analogy to the origin of

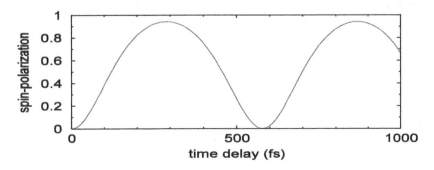

Figure 3.   Spin polarization vs time delay (fs). Figure courtesy of Bouchene et al. (2001).

41

the intrinsic spin Hall effect, by studying the ferromagnetic semiconductor (Ga,Mn)As as a material system since its crystal inversion asymmetry allows the measurement of the bare ferromagnetic films, rather than ferromagnetic–paramagnetic heterostructures, thereby eliminating any spin Hall effect contribution. An intuitive picture of the Berry curvature origin of this anti-damping spin–orbit torque as well as its microscopic modeling has been provided. It has been emphasized that the Berry curvature spin–orbit torque is of comparable strength to the spin-Hall-effect-driven anti-damping torque in ferromagnets interfaced with paramagnets with strong intrinsic spin Hall effect.

It can be visualized that the induced spin polarization (s) is coupled to the magnetization (M) by the exchange interaction (E_{ex}), in terms of the exchange coupling constant (J) by:

$$E_{ex} = -J :: sM \tag{8}$$

which leads to a torque exerted on the magnetization M, which may be written as:

$$\tau = -J :: (M \times s) \tag{9}$$

Since M is along the $z - axis$, and s has two nonzero components, s_x and s_y, the torques exerted on the magnetization are:

$$\tau_y = -J :: (M \times s_x) \tag{10}$$

and

$$\tau_x = -J :: (M \times s_y) \tag{11}$$

which tend to rotate magnetization from the orientation normal to the system to the orientation in the system plane. The magnetization rotation can be fully described by knowing (i) the thermally induced spin torque, for arbitrary magnetization orientation; and (ii) all other torques exerted on the magnetization, including the damping torque, and the torque due to anisotropy fields. Thus, it is clear that for maximizing the performance of the spin lasers the designer, in addition to taking the above mentioned points into consideration, has to optimize τ, J, M, Δ^{-1}, and s, which on the experimental feedback is achieved after a number of iterations.

3   CONCLUDING REMARKS

The studies on the spin lasers have picked up recently, because this system carries the great advantages of both lasers and spintronics, a newly evolving field, considered by many as an offshoot of electronics, or even as a new branch of electronics. These characteristics enable the spin lasers to have a number of applications in research work, and other spintronic systems. The researchers are concentrating their efforts on improving the performance of these lasers, and also reducing their size along with improving their configuration, so as to make them more handy and suitable for their use in various systems. The designing of such lasers provides the designers more degrees of freedom. Hence, it may be concluded that the field of spin lasers is evolving fast, and is on a firm footing.

ACKNOWLEDGMENTS

The authors are grateful to Dr. Nand Kishore Garg, Chairman, Maharaja Agrasen Institute of Technology, GGSIP University, Delhi for providing the facilities for carrying out this research work, and also for his moral support. The authors are thankful to Dr. M. L. Goyal, Vice Chairman for encouragement. Thanks are also due to Dr Neelam Sharma, Director, and Dr. V. K. Jain, Deputy Director for their support during the course of the work.

REFERENCES

Bouchene M.A., Zamith Sébastien, & Girard Bertrand, Coherent control of spin–orbit precession with shaped laser pulses, Journal of Physics B: Atomic, Molecular and Optical Physics 34 (2001) 1497.

Chopra Kamal Nain, Analytical Treatment of the Theory and Design Aspects for the Modeling and Optimization of the Efficiency of Spin Lasers, Atti Fond G. Ronchi, ITALY, 72 (2017) 37–47.

Chopra Kamal Nain, Mathematical Aspects of Spin-related Phenomena Models and the Associated Criteria for Spintronics, L. American Journal of Physics E, LAT. AMERICA 8 (2014) 4313-1–4313-6.

Dyakonov M., Spin Physics in Semiconductors, Springer, 2008.

Gahl A., Balle S. & Miguel M. San, Polarization dynamics of optically pumped VCSEL's, IEEE Journal of Quantum Electronics, 35 (1999) 342–351.

Kurebayashi H., Sinova Jairo [...] Jungwirth T., An antidamping spin–orbit torque originating from the Berry curvature, Nature Nanotechnology 9 (2014) 211–217.

Martin-Regalado J., Prati F., Miguel M. San, & Abraham N.B., Polarization properties of vertical-cavity surface-emitting lasers, IEEE Journal of Quantum Electronics, 33 (1997) 765–783.

Nakajima Takashi, Control of the spin polarization of photoelectrons/photoions using short laser pulses, Appl. Phys. Lett. 84 (2004) 3786.

Advances in Optoelectronic Technology and Industry Development – Jose & Ferreira (eds)
© *2020 Taylor & Francis Group, London, ISBN 978-0-367-24634-1*

Field-free orientation dynamics of CO molecule by utilizing two dual-color shaped laser pulses and lower intensity of THz laser pulse

W.S. Zhan, H.C. Tao & S. Wang
Dalian University of Technology, Panjin, Liaoning, China

ABSTRACT: Field-free orientation of CO molecule is studied theoretically by combining two dual-color shaped laser pulses with low intensity of THz laser pulse. It is indicated that the molecular orientation can be greatly improved by applying two dual-color shaped laser pulse and lower intensity of THz laser pulse compared with single THz laser pulse. The influence of the electric field amplitude of the two dual-color shaped laser pulses on molecular orientation is discussed. Furthermore, by varying the delay time t_{d1} between the two dual-color shaped laser pulses as well as the delay time t_d between the second dual-color shaped laser pulse and THz laser pulse, the molecular orientation can be changed to some extent. Additionally, it's also shown that the enhancement or suppression of the molecular orientation can be coherently manipulated by changing the center frequency and the carrier envelope phase of the THz laser pulse.

Keywords: molecular orientation, two dual-color shaped laser pulses, THz laser pulse

1 INTRODUCTION

In recent years, the molecular alignment and orientation have aroused much interest from many researchers because of their widespread applications in multiphoton ionization (Holmegaard et al., 2010), high-order harmonic generation (Kanai et al., 2005), as well as photoelectron angular distribution (Holmegaard et al., 2010). Alignment refers that the molecular axis is parallel to the field polarization vector, while the orientation with a special "head-versus-tail" order is much more difficult to realize.

Along with the development of THz technology, the researches have achieved better orientation degree via utilizing the THz laser Half-Cycle Pulse (HCP) (Matos-Abiague & Berakdar, 2003) and THz laser Few-Cycle Pulse (FCP) (Qin et al., 2012). However, it is difficult to generate the intense THz laser pulse in the experiment (Backus et al., 1998). In this paper, we propose a scheme to achieve a high-efficiency field-free orientation of CO molecules by using two dual-color shaped laser pulses and a lower intensity of THz laser pulse. The applied intensity of THz laser pulse is 10^8 V/m, which is easier to generate in the experiment.

2 THEORETICAL METHOD

In this scheme, two dual-color shaped laser pulses and THz laser pulse are applied to improve the molecular orientation. And two dual-color shaped laser pulses can be expressed as (Liu et al., 2013):

$$E_{STRT}(t) = E_i(t)[\cos \omega_i(t-t_i) + \cos 2\omega_i(t-t_i)], i = 1, 2 \tag{1}$$

with

$$E_i(t) = E_{0i}\exp\left[-\frac{(t-t_i)^2}{2\sigma_i^2}\right]\{[\sigma_i = \sigma_{ri}(t \leq 0), \sigma_i = \sigma_{fi}(t>0), \sigma_{ri} \gg T_{rot} \gg \sigma_{fi}], i = 1, 2\} \tag{2}$$

where E_{0i} = the amplitude of dual-color shaped laser pulse; ω_i = fundamental frequency and $2\omega_i$ = the second harmonic frequency; t_i = the center time; σ_i = the duration, σ_{ri} = the rising time and σ_{ri} = the falling time; T_{rot} = the rotational period of the CO molecule.

The THz laser pulse is given by

$$E_{THz}(t) = E_3(t)cos[\omega_{THz}(t-t_3) + \varphi] \tag{3}$$

with

$$E_3(t) = E_{03}\exp\left[-2\ln 2(t-t_3)^2/\tau^2\right] \tag{4}$$

where E_{03} = the amplitude of THz laser pulse; ω_{THz} = the center frequency, τ = the Full Width at Half Maximum (FWHM); t_3 = the center time; φ = the Carrier Envelope Phase (CEP). The delay time t_{d1} (between first and second dual-color shaped laser pulse) denotes $t_{d1} = t_2-t_1$ and t_d (between second dual-color shaped laser pulse and THz laser pulse) is calculated by $t_d = t_3-t_2$.

In the rigid rotor approximation, the total Hamiltonian of the molecule interacting with the combination of two dual-color laser pulses and THz laser pulse can be given by (Li et al., 2013)

$$\hat{H}(t) = B_e\hat{J}^2 - \mu E_{THz}(t)\cos\theta - \frac{1}{2}\left[(\alpha_\parallel - \alpha_\perp)\cos^2\theta + \alpha_\perp\right]E_{STRT}^2(t)$$
$$- \frac{1}{6}\left[(\beta_\parallel - \beta_\perp)\cos^3\theta + 3\beta_\perp\cos\theta\right]E_{STRT}^3(t) \tag{5}$$

The molecular orientation degree is defined as

$$\langle\cos\theta\rangle = \text{Tr}\{\cos\theta.\hat{p}(t)\} \tag{6}$$

The time evolution of the density operator is calculated by the quantum Liouville equation

$$\frac{d\hat{\rho}(t)}{dt} = -\frac{i}{\hbar}[\hat{H}, \hat{\rho}(t)] \tag{7}$$

The density operator can be expanded in the eigenstates of the rigid rotor Hamiltonian as

$$\hat{\rho}(t) = \sum_{J,M,J',M'}\rho_{J,M,J',M'}(t)|JM\rangle\langle J'M'| \tag{8}$$

The initial density operator satisfies the temperature-dependent Boltzmann distribution:

$$\rho_0(t) = \frac{1}{Z}\sum_{J=0}^{\infty}\sum_{M=-J}^{M=J}|JM\rangle\langle J'M'|\exp\left(\frac{-B_eJ(J+1)}{k_BT}\right) \tag{9}$$

45

3   RESULTS AND DISCUSSIONS

In our calculation, the CO molecule is used as a model sample. The molecular parameters used in our calculation are chosen as follows (Liu, Y et al. 2013): B_e = 1.93 cm^{-1}, μ = 0.112 D, $\alpha_\parallel$ = 2.294 Å^3, $\alpha_\perp$ = 1.77 Å^3, $\beta_\parallel$ = 2.748 × 10^9 Å^5, $\beta_\perp$ = 4.994×10^8 Å^5. We supposed the polarization axis of the laser pulse as the laboratory-fixed Z axis. The degree of molecular orientation is calculated by the ensemble average of cos θ, where θ is the angle between the laboratory-fixed Z axis and the molecule-fixed Z axis. The definition of the positive orientation is that carbon atom directed the positive direction of the laboratory-fixed Z axis while the negative orientation corresponds to carbon atom directed the negative direction.

Figure 1 shows the time evolution of the molecular orientation manipulated by single THz laser pulse and two dual-color shaped laser pulses with lower intensity of THz laser pulse. The parameters in Figure 1a are taken to be: E_{03} = 1 MV/cm, ω_{THz} = 0.1 THz, τ = 0.5 ps, t_3 = 0 ps, φ = 0 and T = 0 K. The parameters in Figure 1b are taken to be: E_{01} = E_{02} = 5 × 10^9 V/m, σ_{r1} = σ_{r2} = 20 ps, σ_{f1} = σ_{f2} = 0.4 ps, t_1 = −0.95T_{rot}, t_2 = 0 ps, ω_1 = ω_2 = 12500 cm^{-1}, E_{03} = 1 MV/cm, τ = 0.5 ps, t_3 = 0.25 T_{rot}, ω_{THz} = 0.8 THz, φ = 0.6π and T = 0 K. The maximal orientation degree in Figure 1a is quite low, which is less than 0.2. Compared with the single THz laser pulse, the degree of molecular orientation has been greatly improved by combining two dual-color shaped laser pulses with low intensity of THz laser pulse in Figure 1b, which is increased to 0.78. In the following, the influences of laser pulse parameters on the molecular orientation are investigated.

At the beginning, the influence of the delay time t_{d1} between first and second dual-color shaped laser pulse has been analyzed in Figure 2. The parameters are taken to be: E_{01} = E_{02} = 5 × 10^9 V/m, σ_{r1} = σ_{r2} = 20 ps, σ_{f1} = σ_{f2} = 0.4 ps, t_2 = 0 ps, ω_1 = ω_2 = 12500 cm^{-1}, E_{03} = 1 MV/cm, τ = 0.5 ps, t_3 = 0.25 T_{rot}, ω_{THz} = 0.8 THz, φ = 0.6 π and T = 0 K. As shown in Figure 2, the largest degree of the molecular orientation can be obtained at t_{d1} = 0.95T_{rot}. The maximum molecular orientation is 0.613. And with the increase of delay time, the degree of orientation changes obviously. Therefore, varying the delay time t_{d1} can be utilized to enhance or suppress the molecular orientation. In subsequent studies, the delay time t_{d1} between two dual-color shaped laser pulses is set as 0.95 T_{rot}.

The maximal orientation degree as a function of the delay time t_d between second dual-color shaped laser pulse and THz laser pulse is shown in Figure 3. And Figure 3 displays that the highest maximum orientation appears at t_d = 0.25T_{rot} (<cos θ>$_{max}$ = 0.621). Thus, the molecular orientation degrees can be changed to some extent by optimizing the delay time t_d. The optimized delay time in the later calculation is chosen as t_d = 0.25 T_{rot}.

Figure 4 shows the maximal molecular orientation degrees versus the intensity of the two dual-color shaped laser pulses. When the intensity varies from 2.5 × 10^9 to 5.5 × 10^9 V/m,

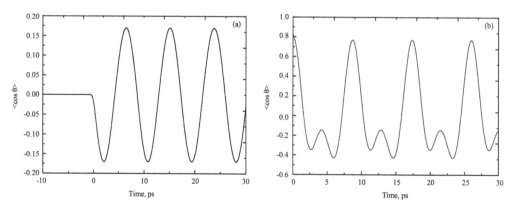

Figure 1.   Time evolution of molecular orientation <cos θ>controlled by: (a) a single THz laser pulse; (b) two dual-color shaped laser pulses with THz laser pulse.

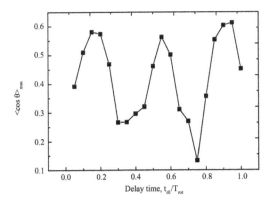

Figure 2. The maximal molecular orientation versus the delay time t_{d1}.

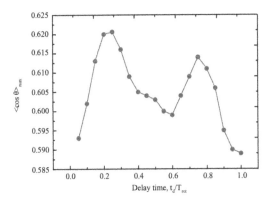

Figure 3. The maximal molecular orientation versus the delay time t_d.

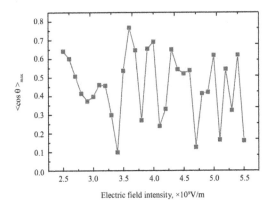

Figure 4. The maximal molecular orientation versus the intensity of dual-color shaped laser pulses.

there is no obvious law that the maximum degree of molecular orientation increases first and then decreases. It is mainly because using two dual-color shaped laser pulses, the electric field intensity has a great influence on the molecular orientation. At the same time, the maximum orientation degree achieved at $E_{01} = E_{02} = 3.6 \times 10^9$ V/m. Thus, the maximum orientation can be effectively controlled by changing the electric field intensity.

Figure 5 depicts the variations of maximum molecular orientation when the central frequency of THz laser pulse changes from 0.05 to 1.2 THz. When the central frequency increases from 0.05 to 0.8 THz, the molecular orientation increases largely. While the central frequency varies from 0.8 to 1.2 THz, the degree of molecular orientation decreases slightly and hardly changes. Therefore, the maximum molecular orientation can be manipulated by optimizing the center frequency of THz laser pulse.

By adjusting the carrier envelope phase, the asymmetry of THz laser pulse will be modified, which will affect the molecular orientation. As can be seen from Figure 6, the molecular orientation reaches its maximum at $\varphi = 0.6\pi$. Moreover, when the carrier envelope phase changes from 0 to 2π, the maximum orientation only varies between 0.763 and 0.770, and the range of change is small. The main reason is that the applied intensity of THz laser pulse is too minor, the influence of changing the phase on the molecular orientation is not obvious. While, the field-free orientation of CO molecule can also be controlled in a certain range by changing the carrier envelope phase.

The above studies of field-free molecular orientation are all considered under the assumption of $T = 0$ K. While in the experiment, the incoherent population will occur in the rotational state. As can be seen from Figure 7, the molecular orientation decreases rapidly with the increase of temperature. When temperature $T = 0, 1, 2$ and 5 K, the maximum orientations of the molecule are 0.770, 0.761, 0.651 and 0.372, respectively. The durations of $T = 0, 1$ and 2 K are 1600 fs, 1500 fs and 1220 fs, respectively. The long duration can be prepared for the generation of high-order harmonics and the detection of electronic dynamics.

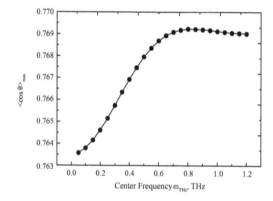

Figure 5.   The maximal molecular orientation versus the center frequency of the THz laser pulse.

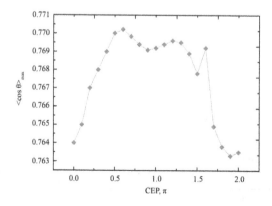

Figure 6.   The maximal molecular orientation degrees versus the CEP of the THz laser pulse.

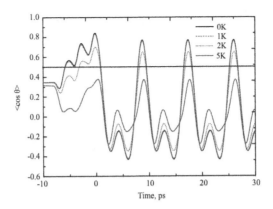

Figure 7.   The maximal molecular orientation degrees versus the rotational temperature.

4   CONCLUSION

A scheme of controlling the field-free orientation of a CO molecule by combining two bi-color shaped laser pulse with lower intensity of THz laser pulse is proposed. Compared with a single THz laser pulse, the molecular orientation has been improved by applying this scheme. And the intensity of THz laser pulse applied is 1 MV/cm, which meets the experimental requirements. The effects of delay time t_{d1} and t_d on molecular orientation are studied theoretically. By adjusting the electric field intensity of the shaped laser pulse, the molecular orientation is improved effectively. The influences of the center frequency and carrier envelope phase of terahertz laser on the orientation is also considered, and the field-free orientation of the CO molecule can also be manipulated to a certain degree. Ultimately, the maximum molecular orientation we obtained was 0.77, which is far larger than the molecular orientation achieved by the single low intensity of THz laser pulse. With the increase of temperature, the degree of molecular orientation decreases rapidly.

REFERENCES

Backus, S et al. 1998. High power ultrafast lasers. Review of Scientific Instruments 69(3): 1207–1223.

Holmegaard, L et al. 2010. Photoelectron angular distributions from strong-field ionization of oriented molecules. Nature Physics 6(6): 428–432.

Kanai, T et al. 2005. Quantum interference during high-order harmonic generation from aligned molecules. Nature 435(7041): 470–474.

Li, H et al. 2011. Orientation dependence of the ionization of CO and NO in an intense femtosecond two-color laser field [J]. Physical Review A 84(4): 043429.

Li, H et al. 2013. Field-free molecular orientation by femtosecond dual-color and single-cycle THz fields. Physical Review A 88(1): 13424.

Liu, Y et al. 2013. Field-free molecular orientation by two-color shaped laser pulse together with time-delayed THz laser pulse. Laser Physics Letters 10(7): 076001.

Matos-Abiague A & Berakdar J. 2003. Sustainable orientation of polar molecules induced by half-cycle pulses. Physical Review A 68(6): 063411.

Qin, C. C et al. 2012. Field-free orientation of CO by a terahertz few-cycle pulse. Physical Review A 85(5): 53415.

Advances in Optoelectronic Technology and Industry Development – Jose & Ferreira (eds)
© 2020 Taylor & Francis Group, London, ISBN 978-0-367-24634-1

54 ps Q-switched microchip laser with a high modulation depth SESAM

L. Gong, H. Zhang, Y.S. Wang, Y. Wang & P.F. Chen
School of Optical and Electronic Information, Huazhong University of Science and Technology (HUST), Wuhan, China,

ABSTRACT: We present a passively Q-switched diode-pumped Nd:YVO$_4$ microchip laser based on a Semiconductor Saturable Absorber Mirror (SESAM) with a high modulation depth of 40%. We obtained 54 ps pulses with 2.9 mW average power at a repetition rate of 550 kHz.

Keywords: Q-switched, microchip laser, SESAM

1 INTRODUCTION

In tens of past years, sub-100-ps pulses were mainly produced by mode-locked lasers (Keller et al., 1996). However, the repetition frequency of mode-locked laser oscillators was too high, usually at the level of tens or hundreds of MHz, and the single pulse energy was low. Consequently, mode-locked lasers required a complex and expensive amplification system to amplify its low pulse energy to high. On the contrary, recently passively Q-switched diode-pumped microchip lasers have attracted attention due to their simple and compact characteristics and low pulse repetition rate. Therefore, Q-switched diode-pumped microchip lasers, as a seed for laser amplifiers, have the advantage of low cost.

Using a saturable absorber such as a Semiconductor Saturable Absorber Mirror (SESAM) as the Q-switched element, sub-100-ps pulses can be generated from Q-switched microchip lasers (Braun et al., 1997; Butler et al., 2012; Mehner et al., 2014). Braun et al. (1997) reported a Q-switched diode-pumped microchip laser with pulse width of 56 ps. Butler et al. (2012) explored the scaling for shorter pulses of Q-switched diode-pumped Nd:YVO$_4$ microchip lasers with SESAM, and obtained pulse width of 22 ps with Nd:YVO$_4$ of 110 um thickness. Mehner et al. (2014) used a very thin Nd:YVO$_4$ crystal of 50 um thickness as the gain medium of Q-switched diode-pumped microchip lasers with SESAM, obtaining 16 ps pulses. The thickness of crystal microchip is one of the key factors for shorter pulses, because the cavity length in the microchip Q-switched laser is approximately equal to the thickness of the crystal microchip. And according to Equation 1, the pulse width, which is related to the cavity length, can be approximately calculated (Braun et al., 1997):

$$\tau_p = 3.52 \frac{T_R}{\Delta R} \tag{1}$$

where T_R is the cavity round trip time, and ΔR is the modulation depth of the absorber. As a consequence, a crystal microchip, as thin as 50 um, has been used in recent studies for generating shortest pulses from Q-switched Nd:YVO$_4$ microchip lasers (Mehner et al., 2014). In fact, a thin crystal microchip is hard to fabricate and to handle. With these concerns, a SESAM with high modulation depth, corresponding to high loss coefficient, is expected to be used in Q-switched Nd:YVO$_4$ microchip lasers for generating short pulses.

In this paper, we use a SESAM with modulation depth of 40% to Q-switch a diode-pumped 300 um thickness Nd:YVO$_4$ microchip laser. As far as we know, a SESAM with such high modulation depth is used in a diode-pumped Q-switched Nd:YVO$_4$ microchip laser for the first time. And we obtain 54 ps Q-switched pulses with 2.9 mW average power at a repetition rate of 550 kHz.

2 EXPERIMENTAL SETUP

Our experimental setup is shown in Figure 1. The pump source is a 2 W fiber coupled laser diode, emitting at wavelength of 808 nm. The core diameter of coupled optical fibers is 105 um and the numerical aperture is 0.15. After collimation by a lens with a focal length of 76 mm and focused by a lens with a focal length of 25.4 mm, the bump beam delivered from the fiber is transmitted to the gain medium at a diameter of 35 um approximately.

The gain medium is a 300 um thick and 3% doped Nd:YVO$_4$ microchip. Such high doping concentration makes it possible to absorb more pump light in a very thin crystal. The upper surface of the microchip is antireflective-coated for wavelength of 808 nm and partial reflective of 80% for wavelength of 1064 nm, which acts as an output coupler of transmission of 20% for laser wavelength of 1064 nm. Therefore, other devices specifically used as the output coupler are not needed, which makes the structure of the Q-switched Nd:YVO$_4$ microchip laser more compact. The lower surface of the Nd:YVO$_4$ microchip is antireflective-coated for laser wavelength of 1064 nm and highly reflective for wavelength of 808 nm, for which the SESAM can be protected and the gain medium can absorb the bump light twice. The SESAM (BATOP GmbH) soldered on a copper heat sink has a relaxation time of 12 ps, non-saturable loss of 10%, saturation fluence of 19 uJ/cm^2 and a modulation depth of 40% at 1064 nm. In this laser, SESAM acts as a highly reflective end mirror and a saturable absorber of the laser resonator. So the Q-switched Nd:YVO$_4$ microchip laser resonator consisted of only the 300 um thick Nd:YVO$_4$ microchip and the SESAM, which makes the resonator very simple and compact. And the lower surface of the microchip is stuck on the SESAM. Light entering the SESAM is only a few microns thick. So the length of the resonator is approximately the thickness of the crystal microchip, for 300 um thick crystal microchip (Spühler et al., 2001). A 45-degree dichroic mirror, antireflective-coated for wavelength of 808 nm and highly reflective for laser

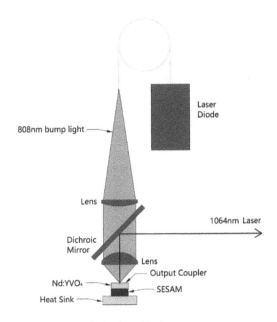

Figure 1.   Experimental setup of Q-switched microchip laser.

wavelength of 1064 nm is placed between two lenses to separate bump beam and the output laser delivered from the laser resonator. Both lenses are antireflective-coated for wavelength of 808 nm and laser wavelength of 1064 nm, which makes the incident bump power in the microchip and the output Q-switched pulses from the laser less loss. The copper heat sink is mounted on a manual displacement table, which makes it possible to adjust the position of the gain medium along the propagation direction of the pump light near the focus of the pump light.

3 RESULTS

With careful adjustment, the diode-pumped microchip laser operated successfully in Q-switched state. With the output coupler of transmission of 20% for laser wavelength of 1064 nm, we obtained 2.9 mW of average power from the Q-switched Nd:YVO$_4$ microchip laser at 0.77 W of bump power. And the repetition rate of the Q-switched pulses was 550 kHz, corresponding to 5 nJ of single pulse energy from the laser. In our experiment, we used an autocorrelator (FEMTO-CHROME, FR-103WS) to measure the autocorrelation curve of the pluses, as shown in Figure 2. Assuming sech^2 fitting, the pulse width was 54 ps. According to theoretical analysis based on Equation 1 proposed in Braun et al. (1997), the theoretical pulse width is 34 ps. Therefore, the experimental is larger than theoretical values. The possible sources of errors between theory and experiment are that the parameters of components given to us that make up the Q-switched Nd:YVO$_4$ microchip laser are inaccurate, measurement errors of the autocorrelator, and theoretical model errors. However, the specific errors need to be further explored in the future.

In order to measure beam quality of the laser at 0.77 W of bump power, we focused the output beam from the Q-switched Nd:YVO$_4$ microchip laser with a lens, and then measured the spot size of the focused beam at different positions along the propagation direction with a CCD. We derived the spot size from the spot measured by the CCD. We used hyperbola to fit the spot size of the output beam of the Q-switched microchip laser at different positions, as shown in Figure 3. And then calculated the beam quality factor of $M_x^2 = 1.20$ in the horizontal direction and of $M_y^2 = 1.22$ in the vertical direction, respectively. The beam quality factors are almost equal in the two orthogonal directions. Because the pump beam delivered from the coupled optical fibers is circularly symmetrical and the diffraction loss of the oscillating laser is mainly determined by the pump spot in the cavity. As a consequence, the beam quality of the Q-switched Nd:YVO$_4$ microchip laser at 0.77 W of bump power is very good. This means that it is enough to dissipate heat from the microchip and the thermal effect of the laser is very small in this case. However, at higher power pumping, the thermal effect of the microchip may become serious, if the cooling capacity is insufficient under this condition of natural cooling. Besides, good beam quality of the laser shows that the fundamental transverse mode of laser matches well with the shaping pump beam in the microchip laser resonator. By changing the lens focal length of the pump shaping system, the size of the pump spot in the microchip will be changed. In this way, we can probably optimize the pump beams in the laser resonator to better match the basic laser transverse modes in the future, which can improve the beam quality of the output laser from the diode-pumped Q-switched Nd:YVO$_4$ microchip laser.

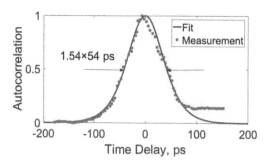

Figure 2.   Autocorrelation trace, assuming sech^2 fit.

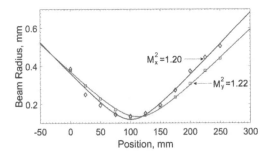

Figure 3. Beam quality of the laser.

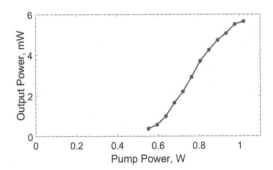

Figure 4. Output power as a function of the pump power.

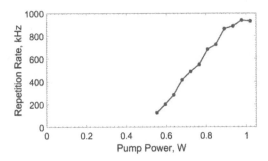

Figure 5. Repetition rate as a function of the pump power.

We used a power meter to measure the output characteristics of the laser with respect to pump power. The relationship between output power and pump power is shown in Figure 4. The pump power threshold is about 0.55 W. The high pump threshold is due to the high transmission of the output coupler, the high non-saturable loss of SESAM and the elliptical polarization of the pump beam delivered from the coupled fiber. In our measurement range, the total slope efficiency of the laser is about 1.1%, which also shows that the whole system has a high loss. And as the pump power increased to a certain extent, the slope efficiency of output power decreased probably resulting from the serious thermal effect at high power bump.

In order to obtain the frequency characteristics of the output pulses of the Q-switched Nd:YVO$_4$ microchip laser, we used a photodiode and an oscilloscope to measure the repetition rate of the pulses at different pump powers. The relationship between the pulse repetition

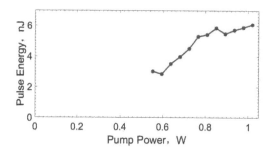

Figure 6.   Single pulse energy as a function of the pump power.

rate of the output laser and the pump power is shown in Figure 5. When the pump power was higher than the threshold, the repetition rate increased from 126 kHz to 936 kHz with the increase of the pump power, which meant that the repetition rate of the pulses from the Q-switched Nd:YVO$_4$ microchip laser can be controlled by changing the pump power in a certain range. The repetition rate of the output pulse of this microchip laser is proportional to the pump power, as predicted by the numerical analysis model in Braun et al. (1997). This means that if we use this diode-pumped Q-switched Nd:YVO$_4$ microchip laser as the seed source of the laser amplifier, we can adjust the repetition frequency of the seed source by changing the pump power of the microchip laser. Besides, according to Equation 1 proposed in Braun et al. (1997), the output pulse width of Q-switched Nd:YVO$_4$ microchip laser with SESAM is only related to the cavity round trip time and the modulation depth of the absorber, so changing the pump power will not cause the change of pulse width. This is very suitable for pulse laser amplification. But as the pump power increases to a certain extent, the repetition rate even decreased. This may be due to that the thermal effect of the microchip will be serious in this case, corresponding to the output power limited at high power pump.

According to the characteristics of output power and repetition rate of pulses output from the Q-switched Nd:YVO$_4$ microchip laser, we can find out the relationship between single pulse energy and pump power, as shown in Figure 6. At pump power ranging from 0.77 W to 1 W, the single pulse energy of the laser is maintained at about 5nJ. According to the numerical analysis in Spühler et al. (2001), the single pulse energy is independent of the pump power. In order to change the single pulse energy, the parameters of the components of the laser resonator must be changed (Spühler et al. 2001).

4   CONCLUSION

We have demonstrated a Nd:YVO$_4$ microchip laser passively Q-switched by a SESAM with modulation as high as 40%. We obtained 54 ps Q-switched pulses by using a 300 um thick Nd:YVO$_4$ microchip as the gain medium of the laser. At 0.77 W of bump power, average output power of the microchip laser was 2.9 mW and the repetition rate of pulses was 550 kHz, corresponding to 5 nJ of single pulse energy, while the beam quality factor M^2 of the laser was 1.20 in the horizontal direction and 1.22 in the vertical direction, respectively. Furthermore, it is expected that a shorter Q-switched pulse can be obtained with a higher modulation depth of SESAM or a thinner crystal in the future. Although the main purpose of our experiment is to measure the pulse width of this Q-switched laser with a high modulation depth SESAM, we also measured the characteristics of the output power and the repetition rate of pulses of the laser. The output power can reach 5.65 mW at pump power of about 1 W. And the repetition rate of the pulses could be varied from 126 kHz to 936 kHz by changing the pump power. But at higher pump power, the output power and the repetition rate of pulses were limited.

REFERENCES

Braun, B. et al. 1997. 56-ps passively Q-switched diode-pumped microchip laser. Optics Letters 22(6): 381–383.

Butler, A. C. et al. 2012. Scaling Q-switched microchip lasers for shortest pulses. Applied Physics B 109 (1): 81–88.

Keller, U. et al. 1996. Semiconductor saturable absorber mirrors (SESAMs) for femtosecond to nanosecond pulse generation in solid-state lasers. IEEE Journal of Selected Topics in Quantum Electronics 2(3): 435–453.

Mehner, E. et al. 2014. Sub-20-ps pulses from a passively Q-switched microchip laser at 1MHz repetition rate. Optics Letters 39(10): 2940–2943.

Spühler, G.J. et al. 2001. Experimentally confirmed design guidelines for passively Q-switched microchip lasers using semiconductor saturable absorbers. Journal of the Optical Society of America B 18(6): 376–388.

Advances in Optoelectronic Technology and Industry Development – Jose & Ferreira (eds)
© 2020 Taylor & Francis Group, London, ISBN 978-0-367-24634-1

Microstructured fiber hydrogen-sensing based on optimized Pd-Ag film

X. Zhou, M. Yang, K.F. Liu, X.Z. Ming & R. Fan
Institute of Mechanical Engineering, Hubei University of Arts and Science, Xiangyang, China

Y.T. Dai
China National Engineering Laboratory for Fiber-Optic Sensing Technology, Wuhan University of Technology, Wuhan, China

ABSTRACT: A novel microstructured Fiber Bragg Grating (FBG) hydrogen sensor was developed by the magnetron sputtering method to prepare alloy films with optimized palladium and silver atom ratios; a femtosecond laser was employed to fabricate spiral microstructures on fiber cladding to improve the flexibility of the fiber. The effects of different palladium and silver atom ratios on the performance of the microstructured FBG hydrogen sensors were investigated. A microstructured fiber sensor with a Pd:Ag atomic ratio of 4:1 was found to give the best hydrogen-sensing performance and offers the prospect of monitoring hydrogen leakages.

Keywords: hydrogen sensor, femtosecond laser, fiber Bragg grating, Pd-Ag

1 INTRODUCTION

Hydrogen is widely used in modern industry as an energy carrier and chemical reaction substance. The monitoring of hydrogen leakage is a key problem because hydrogen gas carries the risk of explosion. Therefore, the study of intrinsically safe hydrogen sensors has attracted widespread attention. Because of Fiber Bragg Gratings' (FBGs) characteristics of intrinsic safety, small size, anti-electromagnetic interference, distributed measurement and easy networking, they are good candidates to apply in the hydrogen-sensing domain (Javahiraly, 2015).

At present, there are many kinds of hydrogen-sensitive materials, such as pure palladium and various palladium alloys. After pure palladium has undergone several cycles of absorbing and releasing hydrogen, a palladium film is prone to cracking or detachment from the surface of a fiber due to phase transition of the Pd film (Ma et al., 2012). In addition, a hydrogen sensor with a pure palladium membrane has a long response time. Therefore, metals such as nickel, silver, and gold are usually incorporated into palladium to form Pd-Ag (Faizal et al. 2015), Pd-Ni (Jiang et al., 2015), and Pd-Au alloy films that can suppress the palladium phase transition (Luna-Moreno & Monzón-Hernández, 2007). Because palladium-silver alloy membranes have better hydrogen selectivity and permeability, it is chosen as a hydrogen-sensitive material by many people. Sharma and Kim (2017) prepared a microelectromechanical (MEMS) hydrogen sensor using a palladium-silver ratio of 77:23. Dai et al. (2013) used a 76:24 ratio of Pd-Ag to prepare a hydrogen sensor. The Pd-Ag film was coated on the FBG surface, which had a diameter of 20.6 µm, and the wavelength shift of the sensor was 40 pm at 4% hydrogen concentration. Jiang et al. (2015) prepared a hydrogen sensor with a palladium-silver ratio of 3:1 composite film on a show side-polished FBG and detected the hydrogen content in oil. Because the ratio of palladium-silver alloy directly affects the sensing sensitivity and response time, it is necessary to explore the influence of different palladium and silver content on the performance of the sensor.

In order to improve the sensitivity of an FBG hydrogen sensor and explore the influence of silver content on sensor performance, a femtosecond laser was used to produce spiral microstructures on FBG cladding, and then different Pd-Ag alloy films were coated on the microstructured FBGs to form hydrogen sensors. The characteristics of the resulting hydrogen sensors were analyzed.

2 EXPERIMENTS

Spiral microstructures were fabricated on the FBG surface using CyberLaser's IFRIT femtosecond laser (780 nm, 180 fs, 1 kHz). Figure 1 shows a schematic view of a spiral microstructured FBG with a length of 10 mm. The laser is focused on the upper surface of the FBG fiber cladding, and then is fed along the straight line. At the same time, the optical fiber is rotated under the grip of a clamp, and thus the production of a spiral microgroove is completed. On the basis of our experience, the requisite processing laser power is $20 \sim 35$ mW, the rotation speed is 12 rpm, the feed rate is 0.72 mm/min, and a microstructured FBG sample with a pitch of 60 μm is prepared. Different atom ratios of Pd-Ag alloy film were sputtered on the samples after microgroove fabrication. As a reference, a non-microstructured standard FBG was also coated. The depth of the deposited film was measured as ~520 nm. The FBG wavelength demodulation was conducted on hydrogen probe samples with an FBG interrogator (SM130, Micron Optics Inc., USA) channel.

3 PRINCIPLES

The center wavelength shift of the FBG fiber is expressed as:

$$\Delta\lambda_{\mathrm{B}} = \lambda_{\mathrm{B}}(1 - p_e)\varepsilon \tag{1}$$

where P_e is the fiber's elastic coefficient, and λ_B is the center wavelength. The relationship between the stress ε and the hydrogen concentration is:

$$\varepsilon = f(H_2) \tag{2}$$

The relationship between the wavelength shift and hydrogen concentration can be obtained by combining Equations 1 and 2:

$$\Delta\lambda_B = \lambda_B(1 - p_e)f(H_2) \tag{3}$$

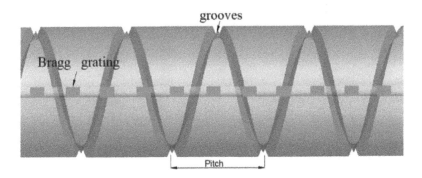

Figure 1.   Schematic diagram of a spiral microstructured fiber.

Because of the increase of the fiber surface area, the amount of Pd-Ag film coated on the fiber will be greater, and the strain generated on the fiber will be larger. Combined with the increased flexibility of the microstructured fiber, the center wavelength shift of the microstructured probe will be greatly improved.

4 DATA AND RESULTS

For Pd-Ag film hydrogen-sensing probes, different Ag content has direct effects on the performance of the sensing probe, including response time, sensitivity and repeatability. Figure 2 presents the wavelength shifts of probes with three kinds of ratios of palladium-silver film in response to a 4% hydrogen concentration. All the probes have the same processing parameters apart from their atomic ratios of Pd-Ag. The wavelength shift of the sample with Pd_4-Ag_1 is about 107 pm at 4% H_2, while for the probe with Pd_2:Ag_1 film it is about 18 pm. Apparently, at the same hydrogen concentration, the probe coated with Pd-Ag film of ratio 4:1 gives the largest wavelength shift, followed by the palladium-silver ratio of 6:1, and the lowest shift is at the ratio of 2:1. Sharma and Kim (2017) reported that a FBG hydrogen gas sensor coated with Pd76-Ag24 film was prepared by a chemical etching method, and the sensitivity of the probe was about 10 pm/% H_2. Coelho et al. (2015) prepared a pure palladium film FBG hydrogen-sensing probe with a sensitivity of 20 pm/% H_2. Compared with these FBG hydrogen-sensing probes, the sensitivity of our spiral microstructured FBG sensor (26.8 pm/% H_2) represents a great improvement.

The comparison of response time of probes with different palladium-silver ratios is shown in Figure 3; the response time and recovery time are defined as that required for 90% of corresponding signal change for absorption and desorption of hydrogen. The response times are 45s, 70s, and 100s for samples Pd_2-Ag_1, Pd_4-Ag_1 and Pd_6-Ag_1, respectively. According to the experimental results, the response time grows with the increasing Ag content. Kim et al. (2008) reported a long-period fiber grating hydrogen sensor coated with 50 nm of pure palladium whose response time was about 8 minutes at 4% hydrogen concentration. Dai et al. (2012) designed a highly sensitive Pd/Ni alloy film FBG probe with a response time of 5~6 minutes at 4% H_2. To sum up, the response time of our FBG probe coated with Pd_4-Ag_1 alloy film represents a great improvement compared with a pure Pd or a Pd-Ni alloy hydrogen sensor. The literature reports that the hydrogen atom permeates the palladium-silver film at the fastest rate (the test ambient temperature is 350°C) when the silver content is 20–23% (Knapton, 1977). It can be simply inferred that the hydrogen atom permeates the palladium-silver film with a silver content of 20%

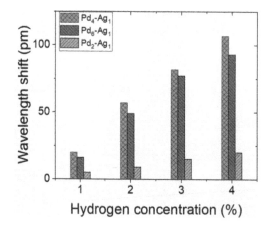

Figure 2. Comparison of hydrogen response of microstructured probes with different palladium-silver ratio alloy films.

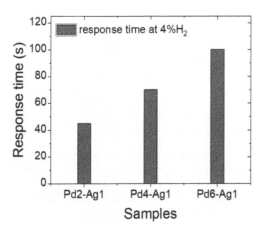

Figure 3. Response time of probes.

faster than a film with a silver content of 25%. Based on the results shown in Figure 3, it can be inferred that the response time of a FBG hydrogen sensor probe coated with palladium-silver film will gradually increase as the silver content gradually decreases. A growing amount of silver can improve the response time of the sensor; however, too much silver in the Pd-Ag alloy will greatly affect the sensitivity of the sensor.

5 CONCLUSIONS

The microstructured fiber hydrogen sensor was fabricated by femtosecond laser and coatings with varying palladium and silver content were applied on the surface of the microstructure. The characteristics of the sensors were studied as the silver content changed. As the silver content increases, the response time of the sensor is shortened, and the optimized atomic ratio of Pd-Ag accelerates hydrogen absorption and penetration. The probe with a film of Pd_4-Ag_1 had the highest sensitivity, and good repeatability and stability. The optimized Pd_4-Ag_1 alloy microstructured FBG hydrogen sensor has good application prospects for monitoring hydrogen leakages.

ACKNOWLEDGMENT

This work was financially supported by the Project of the National Science Foundation of China (NSFC), No. 61475121.

REFERENCES

Coelho, L., De Almeida, J.M.M.M., Santos, J.L. & Viegas, D. (2015). Fiber optic hydrogen sensor based on an etched Bragg grating coated with palladium. *Applied Optics, 54*(35), 10342–10348.

Dai, J., Yang, M., Yu, X., Cao, K. & Liao, J. (2012). Greatly etched fiber Bragg grating hydrogen sensor with Pd/Ni composite film as sensing material. *Sensors and Actuators B: Chemical, 174*, 253–257.

Dai, J., Yang, M., Yu, X. & Lu, H. (2013). Optical hydrogen sensor based on etched fiber Bragg grating sputtered with Pd/Ag composite film. *Optical Fiber Technology, 19*(1), 26–30.

Faizal, H.M., Kawasaki, Y., Yokomori, T. & Ueda, T. (2015). Experimental and theoretical investigation on hydrogen permeation with flat sheet Pd/Ag membrane for hydrogen mixture with various inlet H_2 mole fractions and species. *Separation and Purification Technology, 149*(27), 208–215.

Javahiraly, N. (2015). Review on hydrogen leak detection: Comparison between fiber optic sensors based on different designs with palladium. *Optical Engineering, 54*, 30901–30914.

Jiang, J., Ma, G., Li, C., Song, H., Luo, Y.T. & Wang, H.B. (2015). Highly sensitive dissolved hydrogen sensor based on side-polished fiber Bragg grating. *IEEE Photonics Technology Letters, 27*(13), 1453–1456.

Kim, Y.H, Kim, M.J., Park, M., Jang, J.H., Lee, B.H. & Kim, K.T. (2008). Hydrogen sensor used on a palladium-coated long-period fiber grating pair. *Journal of the Optical Society of Korea, 12*(4), 221–225.

Knapton, A.G. (1977). Palladium alloys for hydrogen diffusion membranes. *Platinum Metals Review, 21*, 44–50.

Luna-Moreno, D. & Monzón-Hernández, D. (2007). Effect of the Pd–Au thin film thickness uniformity on the performance of an optical fiber hydrogen sensor. *Applied Surface Science, 253*(21), 8615–8619.

Ma, G., Li, C., Luo, Y., Mu, R. & Wang, L. (2012). High sensitive and reliable fiber Bragg grating hydrogen sensor for fault detection of power transformer. *Sensors and Actuators B: Chemical, 169*, 195–198.

Sharma, B. & Kim, J.S. (2017). Pd/Ag alloy as an application for hydrogen sensing. *International Journal of Hydrogen Energy, 40*, 25446–25452.

Advances in Optoelectronic Technology and Industry Development – Jose & Ferreira (eds)
© *2020 Taylor & Francis Group, London, ISBN 978-0-367-24634-1*

Synchronous photoelectric scanning imaging in underwater scattering environments

X.Y. Song

Ministry of Education Key Laboratory of Cognitive Radio and Information Processing, Guilin University of Electronic Technology, Guilin, China

Z.Y. He & Y. Huang

Wireless Broadband and Signal Processing Guangxi Key Laboratory, Guilin University of Electronic Technology, Guilin, China

ABSTRACT: Optical imaging is an intuitive method to detect and observe the underwater targets in marine exploration. However, due to the severe effects of light scattering and absorption, especially effected by the back-scattered light, the image quality is dramatically degraded. Based on the transmission behavior of the scattered light, this paper developed a synchronous photoelectric scanning imaging technique. By replacing the mechanical control method with photoelectric scanning, we overcome the implementation problems of mechanical scanning and light source image acquisition devices, finally, alleviate low image quality phenomenon caused by scattering effects. Furthermore, we designed a suitable imaging system and demonstrate it through water tank testing. The experimental results show that the designed scanned based imaging system has higher imaging quality than the non-scanned imaging system.

1 INSTRUCTIONS

Underwater imaging is an important means of seabed detection, it is also a commonly used method for underwater reconnaissance, environmental monitoring, salvage and rescue (Pooja, 2014). Compared with air media, due to the stronger light absorption and scattering characteristics of aqueous media, optical signal attenuation is usually faster, and its image contrast is lower, especially in the case of backscattering, the imaging quality drops dramatically. Although high-power artificial illumination sources and high-sensitivity image acquisition devices can be used to improve image quality, image signal-to-noise ratio is not guaranteed (Jules, 2015). Therefore, under the influence of underwater absorption and scattering, the attenuation of the optical signal intensity is obvious. How to reduce the scattering and even overcome its influence is still a difficult problem of underwater imaging technology.

Methods for solving scattering problems can be divided into three categories: range gating (Yu, 2014), polarized light filtering (Liang, 2017), light source with camera separation, and laser line scan (LLS) imaging system (Shirron & Giddings, 2006). The laser range gated imaging system can effectively improve the imaging quality by shielding backscattering, but its high economic cost limits its application and promotion. Polarized light imaging requires a good understanding of the target and background polarization characteristics, but underwater complex environments often make the signals received by the imaging detectors incapable of accurately characterizing the target and background polarization characteristics, and therefore, without the typical target optical polarization scattering model, this technology is difficult to meet imaging needs. Compared to the first two technologies, LLS can suppress the backscattering effect of water better. Theoretically, LLS imaging resolution can reach the diffraction limit, and the panoramic deep imaging capability is available (Jaffe, 2005).

During the last ten years, enormous efforts have been undertaken to overcome the light attenuation and diffusion owing to the optical absorption and scattering nature of seawater in LLS way. PMT (Dalgleish, 2009; Ouyang, 2013) or linear charge-coupled devices (CCD) (Moore, 2000) were always used as the receiver. The common method to improve the quality of synchronous scanning imaging is to use high-sensitivity detector to realize the tracking and receiving of reflected light in small field of view, however, they are quite complicated and difficult to implement (Bing, 2009, 2014). This prompted us to think about what the key factors in implementing the LLS imaging system is. Nowadays, 2-D array imager sensors with high performance, such as CCD and complementary metal oxide-semiconductor (CMOS) imagers equipped with their mature optics, give one the impression of snap-shot of scanning image. We realize that the key in implementing LLS is how to easily implement the synchronous control of scanning beam and receiving line of sight. Different from the previous LLS systems, which always have mechanical complexity and high imaging design for the receiver, our previous work (He, 2017) shows that a LLS system can be greatly simplified for implementation by using an industrial CMOS camera, and this technique can be easily applied to the undersea imaging except for the higher power lasers needed and waterproof system installation onto a loading vehicle. Thus, in this paper, we simplify the hardware setup of the underwater imaging system with only one CMOS camera and five group of LEDs. The comparisons of direct illumination imaging of the red-light source and our imaging technique show that our imaging system has greater stability, reliability and scanning accuracy.

2 EXPERIMENTAL SETUP

The underwater imaging experiments were carried out in a black tank of size 0.35 m × X 0.35 m × 1.2 m (as shown in Figure 1) filled with 0.33 m depth of water by addition of milk. Attenuation length is used as the measure of the water scattering.

Figure 1.   Experiment setup with tank, camera and laser array.

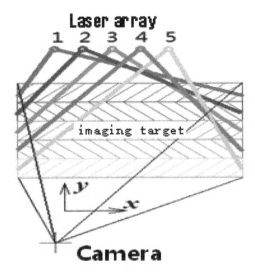

Figure 2.   Top view of laser sources, imaging target and camera. The imaging target is divided into five subareas.

2.1 *Active illumination design*

In this paper, we use a laser array (green 532 nm laser module) as the illumination. The reason is LD array has many advantages such as high luminous efficiency, high reliability, low cost, all solid state, simple driving and small volume and weight. As shown in Figure 2, we align each array element in the LD array with the imaging target separately and use the continuous flash local illumination scanning method to achieve global synchronous scanning imaging at a frame rate of snap-shot imaging in a single frame.

2.2 *Optical receiver*

CCD and CMOS are two types of commonly used imaging devices. The CCD has a shift register charge-coupled unit that corresponds one-to-one with its pixels, and the photo-charges of all the pixels are transferred to the transfer unit at a time, using a global shutter. The two-dimensional photosensitive unit of the CMOS imager has only one row of transmission units, thus the photo-electric image signal can be only taken with the progressive scanning type using a rolling shutter. CCD has advantages over CMOS in terms of integration, frame rate, power consumption and cost, but in terms of processing speed, power consumption, cost and complexity, the rolling shutter is more attractive, hence the CMOS is preferred. In this paper, CMOS is selected as the imaging device, and the QLCAM-CF series camera produced by the Micro vision company is selected as the hardware.

2.3 *Imaging design overview*

Figure 3 shows a typical optical path diagram of a synchronous photoelectric scanning imaging system. The active light source emits continuous narrow beams, it is synchronized with the receiver (a CMOS camera) by the synchronous scanning control module. The advantages of this synchronization are: (1) it can reduce the backscattering volume of the illumination light and improves image quality; (2) it triggers electronic roll shutter exposure through an external trigger port, eliminating the need for mechanical adjustment (translating or rotating) to ensure that the illumination beam is spatially and temporally consistent with the CMOS rolling exposure imaging target area.

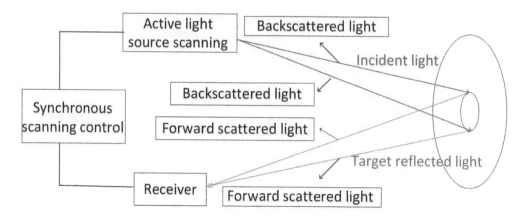

Figure 3. Typical optical path diagram of an LLS system. The red lines are the incident lights, the orange lines are the target reflected lights, and the black lines are the scattered lights.

The light source and the receiver are both narrow field of view, and the overlap between the two fields of view should be very small. Because by reducing the overlapping area of the illuminated water and the receiver, the scattered light entering the receiver and the background noise can be reduced, only backscatter generated by the intersection of the light source beam and the receiving beam will be received, i.e., backscattering in other areas will not enter the receiver, thereby improving the imaging quality. Hence, we place the illumination source and the receiver at an angle to reduce the overlapping area of the illuminated water and the receiving field of view.

Realization process of our synchronous photoelectric scanning imaging: according to the light source layout as shown in Figure 2, laser array sequentially illuminates the corresponding target area of the exposure according to the timing shown in Figure 4 in a frame period, so that the driving pulse can synchronously track the rolling shutter of CMOS imager, and the exposed area of the image sensor is synchronously tracked in space, thereby acquiring an image in which the entire field of view is illuminated.

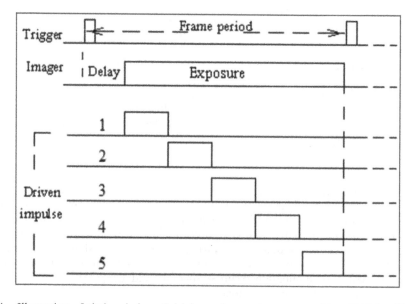

Figure 4. Illustration of timing design of driving pulse synchronous tracking CMOS rolling shutter during single frame period. The trigger method is the same as the one described in [11].

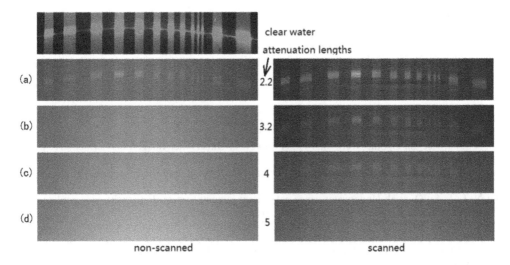

Figure 5.   Image quality comparison of our method and the unscanned method.

3   RESULTS AND ANALYSIS

During the experiment, we placed both the CMOS imaging device and the illumination source on the same side of the imaging target, and the distance between the CMOS and the illumination was greater than one-half of the farthest imaging distance. The reason for this is that it can effectively reduce the impact of backscatter on imaging quality. We simulated different scattering environments of seawater by injecting different amounts of milk into clean water. With the continuous increase of milk, the turbidity of the water medium is increasing, it means the simulated seawater scattering is getting worse (the attenuation lengths is getting longer).

Figure 5 shows the comparison between the imaging results based on the scanning method and the imaging results based on the unscanning method under different attenuation conditions. It can be seen from the changes from (a) to (d) that whether it is our synchronous photoelectric scanning imaging method or non-scanning method, when the attenuation is increased, the clarity of the obtained images is continuously reduced. However, in the same scattering environment, the images obtained by our synchronous photoelectric scanning imaging method are much clearer than the images obtained by the non-scanned method. The experimental results show that our synchronous photoelectric scanning imaging device and the synchronous scanning method proposed in this paper can effectively reduce the influence of scattering on imaging quality.

4   CONCLUSIONS

As suggested by the scanning way, the previous underwater imaging systems sought for a stable translation of the system over the target for image scan. The mechanical complexity and high precise requirements, as well as the customized optical imaging design for the receiver, make it difficult and expensive to build up even an experimental imaging system. This paper makes full use of the operating features of the CMOS, namely the rolling shutter, and designs an underwater imaging system, which can realize the frame synchronization between the image scan and the laser beam sweep and overcome the shortcomings of mechanically controlled scanning method. Experimental results show that the proposed synchronous photoelectric scanning technology can effectively eliminate the influence of scattering on imaging quality even in a simple and low-cost experimental device and has certain engineering practical value.

ACKNOWLEDGMENTS

This work was funded by the: National Natural Science Foundation of China (No. 61761014); Guangxi University Research Projects Funding (No. KY2016YB160); GUET Excellent Graduate Thesis Program (No. 16YJPYBS02).

REFERENCES

Pooja Sahu, Neelesh Gupta, Neetu Sharma, A Survey on Underwater Image Enhancement Techniques, J. Comp. Appl. 87 (13), 19–23, 2014.

Jules S. Jaffe, Underwater Optical Imaging: The Past, the Present, and the Prospects, IEEE J. Oceanic Engin. 40(3), 683–700, 2015.

Yu N, Li L, Su Q, et al. Underwater range-gated laser imaging system design with video enhancement processing [C] International Symposium on Instrumentation & Measurement. 2014.

Liang J, Ju H, Zhang W, et al. Review of Optical Polarimetric Dehazing Technique [J]. Guangxue Xuebao/Acta Optica Sinica, 2017, 37(4).

Shirron J.J, Giddings T.E. A Model for The Simulation of a Pulsed Laser Line Scan System [C] Oceans. IEEE Xplore, 2006.

Jaffe J.S. Performance bounds on synchronous laser line scan systems [C] Oceans. IEEE, 2005.

Dalgleish F.R, Hou W, Caimi F.M, et al. Improved LLS imaging performance in scattering-dominant waters [J]. Proceedings of SPIE - The International Society for Optical Engineering, 2009, 7317:73170E.

Ouyang B, Dalgleish F, Vuorenkoski A, et al. Visualization and Image Enhancement for Multistatic Underwater Laser Line Scan System Using Image-Based Rendering [J]. IEEE Journal of Oceanic Engineering, 2013, 38(3): 566–580.

Moore K.D, Jaffe J.S, Ochoa B.L. Development of a New Underwater Bathymetric Laser Imaging System: L-Bath [J]. Journal of Atmospheric & Oceanic Technology, 2000, 17(8):1106–1117.

Bing Zheng, Bo Liu, Hao Zhang, T. Aaron Gulliver, A Laser Digital Scanning Grid Approach to Three-Dimensional Real-Time Detection of Underwater Targets, 2009.

Bing Ouyang, Weilin Hou, et al., Experimental studies of the compressive line sensing underwater serial imaging system, Ocean Sensing and Monitoring, Proc. of SPIE, 9111, 91110M, 2014.

He Z, Luo M, Song X, et al. Laser line scan underwater imaging by complementary metal-oxide-semiconductor camera [J]. Optical Engineering, 2017, 56(12): 1.

Advances in Optoelectronic Technology and Industry Development – Jose & Ferreira (eds)
© 2020 Taylor & Francis Group, London, ISBN 978-0-367-24634-1

Influence analysis of mixing efficiency of optical heterodyne detection of partially coherent light

Ren Jianying
Graduate School, Space Engineering University, PLA Strategic Support Force, Beijing, China

Sun Huayan, Zhang Laixian & Zhao Yanzhong
Department of Electronic and Optical Engineering, Space Engineering University, PLA Strategic Support Force, Beijing, China

ABSTRACT: Mixing efficiency is an important indicator of laser heterodyne detection systems that directly reflects sensitivity. On the basis of mixing efficiency theory combined with a partially Coherent Gaussian–Schell field model, the expression of mixing efficiency of partially coherent light is derived, and the relationship between spatial mismatch angle, spatial coherence length, receiving radius, and mixing efficiency is obtained. Results from the numerical analysis show that increasing the length of spatial coherence and receiving aperture can improve mixing efficiency but will lead to the reduction of the receiving field of view. When, the laser heterodyne detection system can obtain ideal mixing efficiency and field of view.

1 INTRODUCTION

Among long-distance weak signal detection processes, heterodyne detection has the highest signal-to-noise ratio, and its sensitivity is 20 dB higher than that of direct detection (Das et al., 1997). Heterodyne detection technology requires the mixing of echo signal light and local oscillator light at the receiving surface. Hence, the spatial matching of echo signal light and local oscillator light is strictly required. Mixing efficiency measures the important parameters of heterodyne detection performance. Factors affecting mixing efficiency mainly include spatial matching; polarization state matching; and phase, frequency, and coherence lengths of light field.

Research on mixing efficiency is extensive. Mixing efficiency is studied mainly on the basis of the parameters of the receiving planar light field and spatial position. China Southern Airlines studied the effects of spot size and optical axis deflection on mixing efficiency under a Gaussian beam model. Mixing efficiency was the highest when the ratio of spot size to target surface size was 0.64 (Nan et al., 2017). Li Xiangyang and Han Dong studied the influence of waist radius on heterodyne efficiency on the basis of a Gaussian beam model (Li et al., 2015; Han et al., 2016). Xiang Jinsong studied the influence of beam waist radius on heterodyne efficiency with an Airy disk and a Gaussian model and analyzed the heterodyne efficiency loss caused by aberrations such as defocus, spherical aberration, coma, and astigmatism (Xiang et al., 2009). Liu Hongzhan et al. studied the mixing efficiency of different light field distributions. The analysis results showed that the Airy+Airy and Gauss+Gauss light fields have the best mixing efficiency (Liu et al., 2011).

Wu et al. concluded that "partially coherent light is less affected by turbulence than completely coherent light" by studying the coherent Gaussian–Schell model (GSM) beam in the atmosphere (Jian et al.). These research results focus on partially coherent light, which has become a hot research topic. Ji Xiaoling et al. studied the broadening characteristics of partially coherent hyperbolic cosine Gaussian beams in turbulent transmission (Ji et al., 2006). KeXizheng and Wang Yuting studied the beam broadening characteristics of partially coherent light under different turbulence intensities, transmission distances, and coherence lengths on the basis of the GSM (Ke et al., 2014; Ke et al., 2015). Chen and Yu et al. studied the beam drift characteristics of partially coherent beams in atmospheric turbulent transport (Chen et al., 2013; Yu et al., 2012). Li Lili studied the carrier-to-noise ratio of partially coherent optical laser radar on the basis of the GSM (Pu et al., 2011). The aforementioned studies on partially coherent light mainly involved beam expansion and beam drift in atmospheric turbulence and carrier-to-noise ratio analysis. Hence, the mixing efficiency of partial coherent light in heterodyne detection still needs to be studied in depth.

On the basis of the principle of mixing efficiency and coherent optical transmission theory, the expression of mixing efficiency of partially coherent light is derived theoretically. The mixing efficiency of heterodyne detection is analyzed in terms of spatial mismatch angle, spatial coherence length, and receiving aperture radius of the received light field. Theoretical parameters and numerical analysis are used to obtain the optimal parameters of the optical heterodyne detection system for partially coherent light.

2 HETERODYNE COHERENT DETECTION AND MIXING THEORY ANALYSIS

In a heterodyne coherent detection system, the local oscillator and echo signal light can be expressed as

$$\begin{cases} E_o(\mathbf{r}, t) = U_o(\mathbf{r}) \exp[i(\omega_o t + \varphi_o(\mathbf{r}))] \\ E_s(\mathbf{r}, t) = U_s(\mathbf{r}) \exp[i(\omega_s t + \varphi_s(\mathbf{r}))] \end{cases}. \tag{1}$$

In the formula, $U_o(\mathbf{r})$ and $U_s(\mathbf{r})$ represent the amplitudes of the local oscillator and echo signal light, respectively. ω_o and ω_s represent the angular frequencies of the local oscillator and echo signal light, respectively. $\varphi_o(\mathbf{r})$ and $\varphi_s(\mathbf{r})$ represent the phase of the local oscillator and echo signal light and the spatial position, respectively. $\mathbf{r} = (r, \theta)$ represents the coordinates on the receiving plane of the receiver.

Assuming that the polarization directions of the local oscillator and echo signal light are similar, according to the square rate characteristic of the photodetector, the output photocurrent can be expressed as

$$i(\mathbf{r}) = \frac{eG\eta}{h\nu} \int\int_D \left(|U_o(\mathbf{r})|^2 + |U_s(\mathbf{r})|^2 \right) d\mathbf{r}$$

$$+ \frac{eG\eta}{h\nu} \int\int_D (|U_o(\mathbf{r})||U_s(\mathbf{r})| \exp[i(\Delta\omega t + \varphi(\mathbf{r}))]) d\mathbf{r} + cc. \tag{2}$$

where e represents the electron charge amount, G represents the detector gain factor, η represents the quantum efficiency, h represents the Planck constant, and ν represents the optical frequency. $\Delta\omega = \omega_o - \omega_s$ represents the difference frequency between the local oscillator and the signal light, and $\varphi(\mathbf{r}) = \varphi_o(\mathbf{r}) - \varphi_s(\mathbf{r})$ represents the spatial phase distortion of the local oscillator and signal light. As long as the difference frequency f_s in the angular frequency $\Delta\omega = 2\pi f_s$ is smaller than the cutoff response frequency f of the detector, the detector has a photocurrent output.

The first term in Equation 2 represents the DC signal, whereas the second term represents the heterodyne AC signal of the detector response.

$$i_{AC}(\mathbf{r}, t) = \frac{eG\eta}{h\nu} \int\int_D U_o(\mathbf{r}) U_s^*(\mathbf{r}) \exp[i(\Delta\omega t + \varphi(\mathbf{r}))] d\mathbf{r} + cc. \tag{3}$$

where $\Delta\omega$ represents the difference frequency between the local oscillator and the echo signal light and $\varphi(\mathbf{r}) = \varphi_o(\mathbf{r}) - \varphi_s(\mathbf{r})$ represents the phase difference between the local oscillator and the echo signal light.

Mixing efficiency is defined as the ratio of total coherent detection power P_{total} to the maximum mixing power that can be obtained when the amplitude and phase of the overlapping plane beams of the receiving plane have the same spatial variation. In the heterodyne detection system, if the heterodyne efficiency is low, then the intermediate frequency signal output is weak, and the signal cannot be processed and recognized. Therefore, studying the mixing efficiency of echo signal light and local oscillator light is of great importance to improve the performance of the detection system (Salem et al., 2004). Mixing efficiency can be expressed as

$$\eta = \frac{\mathrm{Re} \iint\limits_D \Gamma_o(\mathbf{r}_1, \mathbf{r}_2)\Gamma_s^*(\mathbf{r}_1, \mathbf{r}_2) \exp[i(\varphi(\mathbf{r}_1) - \varphi(\mathbf{r}_2))]d^2\mathbf{r}_1 d^2\mathbf{r}_2}{\iint \Gamma_o(\mathbf{r}, \mathbf{r})d\mathbf{r}^2 \iint \Gamma_s(\mathbf{r}, \mathbf{r})d\mathbf{r}^2}. \tag{4}$$

3 ANALYSIS OF OPTICAL MIXING EFFICIENCY FOR PARTIALLY COHERENT LIGHT

In heterodyne laser detection, spatial mismatch angle, optical system aberration, atmospheric turbulence, and spot size affect the mixing efficiency of the local oscillator and echo signal light. The spatial mismatch angle means that the focal plane of the receiving lens coincides with the surface of the detector and that the local oscillator light vertically enters the surface of the detector. The echo signal light is incident on the receiving plane at an angle. The spatial positional relationship between the local oscillator and the echo signal light is shown in Figure 1. The spatial mismatch angle causes the wave front of the mixing beam to tilt, which affects the performance of heterodyne detection.

Gaussian caliber is approximated by using a circular aperture function (Salem et al., 2004). The Gaussian aperture function can be expressed as

$$\Phi(\mathbf{r}) = \exp\left(-\frac{\mathbf{r}^2}{\sigma_d^2}\right), \tag{5}$$

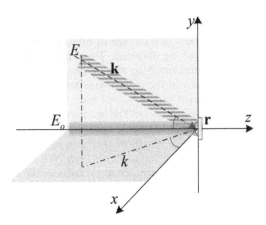

Figure 1.   Schematic diagram of the position of the local light and signal light field at the receiving plane.

where $\sigma_d = \sqrt{d^2/2}$ represents the radius of the hole function. $\varphi(\mathbf{r}) = \mathbf{k} \cdot \mathbf{r} = kr\sin\theta\cos\phi$ represents the phase difference between the local oscillator and the signal light on the receiving plane. Assuming that the local oscillator and echo signal light on the receiving plane are partially coherent GSM light, the intensity distribution model is

$$I(\mathbf{r}_1, \mathbf{r}_2, 0) = I_n \exp\left(-\frac{\mathbf{r}_1^2 + \mathbf{r}_2^2}{2w_n^2}\right). \tag{6}$$

The cross-correlation function of the local oscillator and echo signal light at the receiving plane ($z = 0$) is defined as

$$\Gamma_n(\mathbf{r}_1, \mathbf{r}_2, 0) = I_n \exp\left(-\frac{\mathbf{r}_1^2 + \mathbf{r}_2^2}{2w_n^2}\right)\exp\left(-\frac{(\mathbf{r}_1 - \mathbf{r}_2)^2}{2\rho_n^2}\right), \tag{7}$$

where $n = o$ or s, w_n represents the spot radius, ρ_n represents the coherence length of the beam, $\rho_n \to \infty$ represents fully coherent light, and $\rho_n \to 0$ represents completely incoherent light. $d^2\mathbf{r} = rdrd\phi$ represents the increment on the receiving plane. The mixing efficiency of the coherent detection of the Gaussian aperture function can be expressed as (Mohamed et al., 2010; Andrews et al., 1988):

$$\eta = \frac{\mathrm{Re}\int_{\phi_2=0}^{2\pi}\int_{\phi_1=0}^{2\pi}\int_{r_2=0}^{d}\int_{r_1=0}^{d}\Gamma(\mathbf{r}_1,\mathbf{r}_2)\Phi(\mathbf{r}_1,\mathbf{r}_2)\Psi(r_1,r_2)r_1r_2dr_1dr_2d\phi_1d\phi_2}{\int_{\phi=0}^{2\pi}\int_{r=0}^{d}\Gamma_o(\mathbf{r},\mathbf{r})\Phi(\mathbf{r})rdrd\varphi\int_{\phi=0}^{2\pi}\int_{r=0}^{d}\Gamma_s(\mathbf{r},\mathbf{r})\Phi(\mathbf{r})rdrd\phi}, \tag{8}$$

where $\Gamma(\mathbf{r}_1, \mathbf{r}_2) = \Gamma_o(\mathbf{r}_1, \mathbf{r}_2)\Gamma_s^*(\mathbf{r}_1, \mathbf{r}_2)$ and $\Psi(r_1, r_2) = \exp[ik(r_1\theta\cos\phi_1 - r_2\theta\cos\phi_2)]$. After mixing the partially mixed light efficiency,

$$\eta = \frac{1}{AR}\left(\frac{1}{w_o^2} + \frac{1}{\sigma_d^2}\right)\left(\frac{1}{w_s^2} + \frac{1}{\sigma_d^2}\right)\exp\left(\frac{k^2\theta^2 + T^2}{4R} - \frac{k^2\theta^2}{4A} + \frac{k\theta T}{2R}\right) \tag{9}$$

where $R = A + B^2/4A$, $T = Bk\theta/2A$, $A = \frac{1}{2w_o^2} + \frac{1}{2w_s^2} + \frac{1}{2\rho_o^2} + \frac{1}{2\rho_s^2} + \frac{1}{\sigma_d^2}$, and $B = \frac{1}{\rho_o^2} + \frac{1}{\rho_s^2}$.

4 DATA ANALYSIS

Through the numerical analysis, the visual change of mixing efficiency, the spatial mismatch angle, and the receiving aperture can be obtained. Thus, the optimal solution of the parameters of the heterodyne detection and receiving system can be derived accordingly. The main parameters of the numerical analysis of the heterodyne detection system are as follows: laser wavelength = 532 nm, local oscillator radius $w_o = 4$ mm, signal light radius $w_s = 2$ mm, and signal light spatial coherence length $\rho_s = \infty$. The numerical analysis results are shown in Figure 2. The beam mismatch angles in the figure are 0 rad, 0.0005 rad, 0.001 rad, and 0.005 rad, respectively. The spatial coherence length of the local oscillator is high under the same spatial mismatch angle. Hence, the mixing efficiency is high. When the coherent lengths of the local oscillator are the same, the spatial mismatch angle reduces the mixing efficiency, and the receiving aperture becomes large. Efficiency is highly affected by the mismatch angle.

The signal light radius is 5 cm, the receiving system radius is 5 cm, the spatial coherence length of the signal light is infinite $\rho_s = \infty$, the mismatch angle $\theta = 0$, the ratio of the local oscillator light to the receiving aperture is different, and the mixing efficiency with the local oscillator radius is obtained (Figure 3). Under the condition in which the receiving radius is

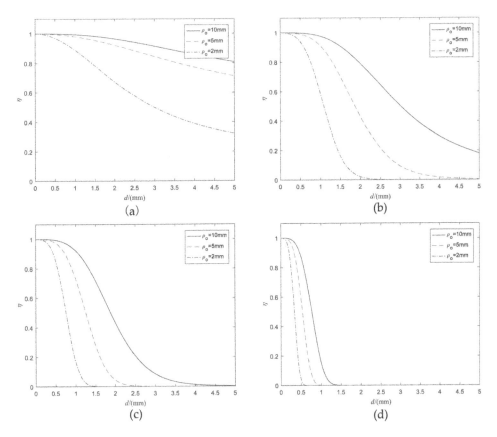

Figure 2. Curve of mixing efficiency variation with receiving aperture under different spatial mismatch angles (a) $\theta = 0$, (b) $\theta = 0.0005$, (c) $\theta = 0.001$, (d) $\theta = 0.005$.

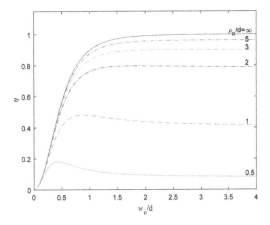

Figure 3. When the ratio of coherence length to receiving aperture is different, mixing efficiency varies with the local oscillator radius.

fixed, the mixing efficiency tends to be stable with the increase of the local oscillator radius. The larger the coherent length of the local oscillator is, the higher the mixing efficiency value is. Figure 4 shows the different local oscillator spot radii. The mixing efficiency varies with the

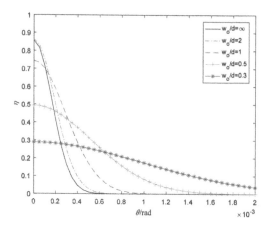

Figure 4. Variation of mixing efficiency with spatial mismatch angle under different local oscillator radii.

spatial mismatch angle. The spatial coherence length of the local oscillator is 5 mm. The mixing efficiency of the local oscillator radius is greater than that of the receiving aperture, and it decreases rapidly with the increase of the mismatch angle. When the ratio is less than 1, the mixing efficiency changes slowly with the mismatch angle. The mixing efficiency when the mismatch angle is less than 0.0002 rad is lower than the mixing efficiency when the ratio is greater than 1. Moreover, the mixing efficiency is greater than the ratio when the mismatch angle is greater than 0.0002 rad at 1 h mixing efficiency. That is, the larger the radius of the local oscillator is, the more heterodyne efficiency is affected by the mismatch angle. The numerical analysis results show that the mixing efficiency and effective field angle can be considered when $w_o/d \approx 0.5$. As shown in Figure 5, under the spatial coherence length of different local oscillators, mixing efficiency varies with the spatial mismatch angle, and the spatial coherence length of the signal light is. Mixing efficiency decreases with the increase of the mismatch angle. Consistent with the previous analysis, the ratio of the spatial coherence length of the local oscillator to the receiving aperture radius is high, thereby making the mixing efficiency high. The effect of the spatial mismatch angle is relatively small. The spatial coherence length can be obtained as $\rho_o/d > 1$ in combination with Figure 3; hence, a relatively good mixing efficiency and field angle range can be obtained.

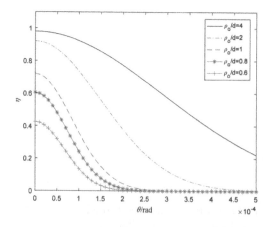

Figure 5. Variation of mixing efficiency with spatial mismatch angle under different coherent lengths of local oscillator.

5 CONCLUSIONS

Mixing efficiency is an important indicator that measures the coherent detection of laser heterodyne. Factors affecting mixing efficiency mainly include the parameters of the local oscillator and signal light, as well as the spatial mismatch angle and receiving aperture. Through the partially coherent GSM, the mixing efficiency expressions of partially coherent light are obtained around the parameters of the local oscillator, signal light field, and mismatch angle. On the basis of this finding, the spatial mismatch angle and space are numerically analyzed, and the effect of coherence length and receiving aperture radius on mixing efficiency is determined. The results show that increasing the spatial coherence length and receiving aperture can improve the mixing efficiency of heterodyne detection, but it will result in a decrease in the receiving field of view. Therefore, for laser heterodyne detection systems, the relationship between the spatial coherence length and the receiving aperture and the angle of view should be fully considered. The analysis results show that $\rho_o/d > 1$ is guaranteed in system design, and the laser heterodyne detection system can obtain the ideal mixing efficiency and field of view.

REFERENCES

Andrews L.C., R.L. Philips. Laser Beam Propagation through Random Media [J]. SPIE Optical Engineering Press, 1988.

Chen C, H. Yang. Temporal spectrum of beam wander for Gaussian Shell-model beams propagating in atmospheric turbulence with finite outer scale [J].Opt Lett, 2013, 38(11): 1887–1889.

Das K.K, Iftekharuddin K.M, Karim M.A. Improved heterodyne mixing efficiency and signal-to noise ratio with an array of hexagonal detectors[J]. Applied Optics, 1997, 36(27):7023–7026.

Han Dong, Liu Yunqing, Zhao Xin, Chu Wei. Effect of Beam Radius on Heterodyne Efficiency for Space Coherent Optical Communication. Journal of Changchun University of Science and Technology, 2016, 39(3): 36–40.

Ji Xiaoling, Huang Taixing, Lv Baida. Spreading of partially coherent cosh Gaussian beams propagating through turbulent atmosphere. ACTA PHYSICA SINICA, 2006, 55(2): 978–982.

Jian, W. Propagation of a Gaussian-Schell Beam Through Turbulent Media. Journal of Modern Optics, 374, 671–684, DOI: 10.1080/09500349014550751.

Ke Xizheng, Han Meimiao, Wang Mingjun. Spreading and wander of partially coherent beam propagating along a Horizontal-Path in the Atmospheric Turbulence. ACTA OPTICA SINICA, 2014, 34(11): 1106003.

Ke Xizheng, Wang Wanting. Expansion and angular spread of partially coherent beam propagating in atmospheric turbulence. Infrared and Laser Engineering, 2015, 44(9).

Li Xiangyang, Ma Zongfeng, Shi Dele. Effect of Gaussian fields distribution on mixing efficiency for coherent detection. Infrared and Laser Engineering, 2015. 44(2): 539–543.

Liu Hongzhan, Ji Yuefeng, Xu Nan, Liu Liren. Effect of Amplitude Profile Difference of Signal and Local Oscillator Wave on Heterodyne Efficiency in the Inter-Satellite Coherent Optical Communication System. ACTA OPTICA SINICA, 2011. 31(10).

Mohamed Salem, Jannick P. Rolland. Heterodyne efficiency of a detection system for partially coherent beams [J]. Optical Society of America, 2010, 27(5): 1111–1119.

Nan Hang, Zhang Peng, Tong Shoufeng et.al. Analysis of the effect of light spot size and ray axis deflection on heterodyne efficiency of space optical hybrid. Infrared and Laser Engineering, 2017.46(4).

Pu Lili, Zhou Yu, Sun Jianfeng, Shen Baoliang, Lu Wei. Receiving characteristics of coherent ladar under partially coherent condition. ACTA OPTICA SINICA, 2011, 31(12).

Salem M, A. Dogariu. Optical heterodyne detection of random electromagnetic beams [J]. J.Mod.Opt., 2004, 51(15): 2305–2313.

Xiang Jinsong, Pan Lechun. Heterodyne Efficiency and the Effects of Aberration for Space Coherent Optical Communication. Opto-Electronic Engineering, 2009, 36(11): 53–57.

Yu S, Z. Chen, T. Wang, et al. Beam wander of electromagnetic Gaussian-Schell model beams propagating in atmospheric turbulence [J]. Appl Opt, 2012, 51(31): 7581–7585.

Advances in Optoelectronic Technology and Industry Development – Jose & Ferreira (eds)
© *2020 Taylor & Francis Group, London, ISBN 978-0-367-24634-1*

Mode competition and cavity tuning characteristics of a new integrated orthogonal polarized He-Ne laser with Y-shaped cavity

Jiabin Chen, Guangzong Xiao & Bin Zhang
Department of Optoelectronic Engineering, College of Advanced Interdisciplinary Studies,
National University of Defense Technology, Changsha, Hunan, P.R. China

ABSTRACT: Aiming at the new integrated orthogonal polarized He-Ne laser with Y-Shaped Cavity, an experimental system for testing cavity tuning characteristic (including the light intensity tuning and frequency difference tuning) is built. By tuning the voltage of the piezoelectric ceramics on the two sub-cavities of S and P, the cavity length of the two sub-cavities is changed to obtain different split frequency differences. In the case of different split frequency differences, the voltage of the piezoelectric ceramic PZT1 on the public cavity mirror is tuned to obtain the light intensity tuning curve of the laser and the corresponding beat frequency variation curve. By tuning the public cavity or S sub-cavity, the double S longitudinal modes are stabilized at both edges of the gain curve, and then the P sub-cavity is continuously tuned to obtain a frequency difference tuning curve. The mechanism of mode competition in the laser is analyzed by using the Lamb semi-classical gas laser theory within the third-order perturbation approximation. The analysis shows that the split frequency difference is the main factor affecting the mode competition. The split frequency difference affects the linear gain and self-saturation effect, mutual saturation effect and the loss of each longitudinal mode. These four factors, which obey the self-consistent equation of light intensity, combine to influence the intensity and competition result of mode competition, thus affecting the change of light intensity. On the basis of the analysis, the longitudinal mode distribution and competition process in each working stage of the laser are theoretically analyzed. The theoretical analysis and interpretation of the intensity tuning curve, the corresponding beat frequency variation and the frequency difference tuning curve obtained by the experiment are carried out. The influencing factors and tuning laws of the light intensity tuning curve are summarized. In the end, the experiment verifies that when the split frequency difference is in the range of 129—1302 MHz, the laser is basically in the working state which the single longitudinal mode pair (including a S longitudinal mode and the neighboring P longitudinal mode) is oscillating.

Keywords: Y-shaped cavity, orthogonal polarized He-Ne laser, light intensity tuning, frequency difference tuning

1 INTRODUCTION

In lasers, mode competition is one of the important physical phenomena, divided into weak, medium and strong (Lamb W E. 1964). It is difficult to observe the strong mode competition in ordinary He-Ne lasers. But the study of mode competition in orthogonal polarized lasers is more effective and convenient. On the one hand, the interval of longitudinal mode frequency is adjustable, which is convenient to study the mode competition and cavity tuning characteristics while the longitudinal mode in different frequency intervals. On the other hand, by separating the two orthogonal polarized longitudinal modes with a polarizing beam splitter, the characteristics and mutual influence of the two longitudinal modes can be studied separately. The splitting of the orthogonal polarized laser frequency by using the birefringence effect was reported in reference

(Zhang Shulian.2005). Based on this, many types of orthogonal polarized lasers were developed to construct a complete academic system of principles, devices, phenomena and applications (Yang Sen, Zhang Shulian.1988; Zhang Shulian, Lu Men, Wu Minxian et al. 1993; Jin Yuye, Zhang Shulian, Li Yan et al. 2001). Reference (Guangzong Xiao. 2011) reports the first orthogonal polarized He-Ne laser with Y-Shaped Cavity which uses a polarizing beam splitter named PBS to achieve the output of the orthogonal polarized longitudinal modes. Based on this, an integrated Y-shaped laser was developed and analyzed (Gong Mengfan. 2015). Foreign scholars such as T. Yoshino had studied orthogonal polarized lasers (T. Yoshino, M. Kawata, B. Qimude et al. 1998; T. Yoshino. 2008).

Cavity tuning, which is the tuning of the cavity length, includes intensity tuning and frequency difference tuning. Light intensity tuning refers to changing the length of the cavity to change the position of the longitudinal mode of the laser on the gain curve. It is a method of studying the model competition by observing the change of light intensity of two orthogonal polarized longitudinal modes. The frequency difference tuning refers to changing the interval of the two orthogonal polarized longitudinal modes continuously by tuning the length of a certain sub-cavity to obtain a curve of the frequency difference as a function of the voltage of the piezoelectric ceramics on the sub-cavity. Reference (Han Yanmei, Zhang Shulian, Li Kelan. 1997) reports the intensity tuning curve of e-light and o-light while the longitudinal modes in different frequency intervals. Reference (Han Yanmei, Zhang Shulian, Li Yan et al. 1998) concludes that the light intensity tuning curve reflects the mutual saturation effect of light intensity by using the theory within the third-order perturbation approximation. The intensity tuning curves and corresponding beat frequency curves of the discrete and integrated lasers with Y-shaped cavity in the case of different split frequency differences were reported in reference (Guangzong Xiao. 2011; Gong Mengfan. 2015), respectively.

This paper is organized as follows. Section 2 introduces the experimental system for testing cavity tuning characteristic. Section 3 summarizes the theoretical analysis of the law of light intensity tuning. Section 4 introduces the experimental results and analysis. Section 5 concludes this paper.

2 EXPERIMENTAL SYSTEM AND METHOD

The experimental system is shown in Figure 1. The device in the red dotted frame is a new integrated orthogonal polarized He-Ne laser with Y-Shaped Cavity (hereinafter referred to as "Y-shaped" laser). The cavity from M1 to PBS is a common section of S-polarized light and P-polarized light (hereinafter referred to as S light and P light), which is called a public cavity. The length is approximately 73mm. The resonant cavity formed by PBS and M3 has only S light oscillation, whose optical cavity length is about 104.8 mm and it is called S sub-cavity. The resonant cavity formed by PBS and M2 has only P light oscillation, which is called P sub-cavity, and the optical cavity length is about 102.9 mm. The PBS is a polarizing beam splitter, which has a reflectivity of more than 99.9% for S light and a transmissivity of more than 99.9% for P light. M1 is the output mirror. M2 and M3 are mirrors with high reflectivity. W is the optical windows with antireflection coating. M4 is the semi-reflecting and semi-transmitting mirror, also called beam splitter. Polarizer is the Polarizer with the polarization angle of 45 degrees, so that the two polarized lights can generate the beat frequency, that is, the frequency difference. D1 and D2 are light intensity detectors. The APD is an avalanche photodiode that is used to acquire the beat frequency signal which is displayed on the spectrum analyzer called SA. M5 is a polarizing cube beam splitting prism for reflecting S light and transmitting P light to achieve splitting. A&F (amplifier and filter) represents amplification filter circuit. DAQ is the data acquisition circuit. PS is the piezoelectric ceramic driving circuit. PZT1 and PZT2 are piezoelectric ceramics, respectively installed on M1, M2. PC is a computer.

In the experiment, after the laser exits from M1, it is divided into two optical paths through M4. APD collects the frequency difference signal on a path, andM5 splits the S light and P light

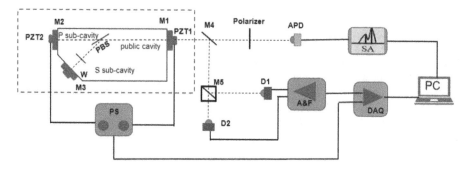

Figure 1.　Cavity tuning characteristic experimental system.

on the other path, on which S light and P light are collected by D1 and D2 respectively. After being amplified by A&F, the light intensity signal is collected by DAQ and sent to the PC.

The process of light intensity tuning is as follows. The LabVIEW software controls the DAQ output to continuously adjust the DC signal. After being amplified by the piezoelectric ceramic drive circuit, it outputs a continuously variable DC voltage of 0V to 300V on the PZT1. Thereby, the continuous change of the length of the public cavity is pushed by PZT1, and the different longitudinal modes are swept on the gain curve to obtain the tuning curve of the light intensity as the function of the voltage of PZT1. By the way, the change of frequency difference is recorded by a spectrum analyzer. Similarly, the frequency difference tuning is to obtain the tuning curve of the frequency difference by changing the voltage of PZT2.

3　THEORETICAL ANALYSIS OF THE LAW OF LIGHT INTENSITY TUNING

In general, in a "Y-shaped" laser, the two adjacent longitudinal modes whose polarization directions are perpendicular to each other have the greatest influence on the light intensity tuning curve. These two longitudinal modes are simply referred to as "S mode" and "P mode", respectively. Regardless of the premise of back scattering, the Lamb semi-classical gas laser theory within the third-order perturbation approximation is used to systematically analyze the influencing factors of the light intensity tuning curve (Xiao Guangzong, Long Xingwu, Zhang Bin, Lu Guangfeng, Zhao Hongchang. 2011). According to Lamb's solution of the light intensity self-consistent equation and the motion matrix of the moving atom's density matrix, the operation formula of the operation formulas of "S mode" and "P mode" in "Y-shaped" laser are obtained as follows (Zhang Shulian. 2005):

$$\dot{I}_S = 2I_S(\alpha_S - \beta_S I_S - \theta_{SP} I_P). \tag{1}$$

$$\dot{I}_P = 2I_P(\alpha_P - \beta_P I_P - \theta_{PS} I_S). \tag{2}$$

Where I_s and I_p are the dimensionless intensity of S light and P light, respectively; α_s and α_p are the linear net gain coefficients of S light and P light, respectively; β_s and β_p are the self-saturation coefficients of S light and P light, respectively; θ_s and θ_p are the mutual saturation coefficients of S light and P light, respectively. Review the knowledge of laser physics (Lu Yaxiong, Yu Xuecai, Zhang Xiaoxia. 2005) and related literature (Gong Mengfan. 2015) to know:

3.1 Linear net gain coefficient

The linear net gain coefficient α is defined as:

$$\alpha_n = g_n - \gamma_n, \quad n = s, p \tag{3}$$

where g_n is the single pass gain coefficient provided by the gain medium, and γ_n is the single pass loss of the longitudinal mode. The definitions are as follows:

$$g_n = L(\omega - \upsilon_n)F_1, \quad n = s, p \tag{4}$$

$$\gamma_n = \upsilon_n/2Q_n, \quad n = s, p \tag{5}$$

where $L(\omega - \upsilon_n)$ is the dimensionless Lorentz function, F_1 is the first order factor. The definitions are as follows:

$$L(\omega - \upsilon_n) = \gamma^2[\gamma^2 + (\omega - \upsilon_n)^2]^{-1}, \quad n = s, p \tag{6}$$

$$F_1 = \frac{1}{2}\upsilon_n D^2 \overline{N}(\varepsilon_0 \hbar \gamma)^{-1}, \quad n = s, p \tag{7}$$

where υ_s and υ_p are the frequency of S light and P light, ω is the center frequency of the gain curve.

According to Equation 4, single pass gain coefficient g_n is mainly related to $L(\omega - \upsilon_n)$, and the smaller $(\omega - \upsilon_n)$ is, the larger g_n is. Therefore, the closer the longitudinal mode is to the center of the gain curve, that is, the smaller $(\omega - \upsilon_n)$ is, the larger the gain coefficient is.

The single pass loss factor γ is related to the geometric deflection loss and diffraction loss of the laser, regardless of the longitudinal mode position. However, during the tuning process, as the piezoelectric ceramic voltage increases, the displacement of the PZT changes nonlinearly, resulting in a nonlinear change in the single path loss factor of each longitudinal mode.

3.2 Self-saturation coefficient

The self-saturation coefficient β is defined as

$$\beta_n = L^2(\omega - \upsilon_n)F_3, \quad n = s, p \tag{8}$$

$$F_3 = \frac{3}{2}\gamma_{ab}F_1\gamma^{-1} \tag{9}$$

Where F_3 is a third-order factor, depending on the degree of loss of the longitudinal mode in the cavity.

According to Equations 8 and 9, the position of the longitudinal mode on the gain curve and the loss in the cavity determine the magnitude of the self-saturation coefficient of the longitudinal mode. The magnitude of the self-saturation coefficient is inversely proportional to the distance from the longitudinal mode to the center frequency of the gain curve and is proportional to the loss in the cavity.

3.3 Mutual saturation coefficient

Mutual saturation coefficient θ can be approximated by the third-order perturbation approximation theory as follows:

$$\theta_{sp} = \theta_{ps} = \theta \tag{10}$$

The definition is

$$\theta = [L(\omega - \Delta\upsilon_1) + L(\Delta\upsilon_2)]F_3 + \frac{1}{2}(\gamma_a\gamma_b\gamma_{ab})F_3 \times$$
$$Re\{[D_a(\upsilon_s - \upsilon_p) + D_b(\upsilon_s - \upsilon_p)] \times [D(\omega - \upsilon_p)N_2/\overline{N} + D(\Delta\upsilon_2)]\} \tag{11}$$

where $\Delta\upsilon_1 = \frac{1}{2}\upsilon_s + \frac{1}{2}\upsilon_p$, $\Delta\upsilon_2 = \frac{1}{2}\upsilon_s - \frac{1}{2}\upsilon_p$. There are two types of frequency-dependent functions in Equation 11. One is a function of the split frequency difference, including the following formula:

$$D_a(\upsilon_s - \upsilon_p)D_b(\upsilon_s - \upsilon_p)D(\Delta\upsilon_2)$$

The other is a function of the difference between the frequency of the two longitudinal modes and the center frequency, including the following formula:

$$L(\omega - \Delta\upsilon_1) D(\omega - \upsilon_p)$$

According to the literature (Xiao Guangzong, Long Xingwu, Zhang Bin, Lu Guangfeng, Zhao Hongchang. 2011), the mutual saturation coefficient θ decreases as split frequency difference increases, and becomes smaller as $(\omega - \Delta\upsilon_1)$ increases.

The difference between the center frequencies of the gain curves of Ne^{20} and Ne^{22} is 875 MHz. Considering the case of filling the Ne double isotope, when the two split frequencies are symmetric with the center frequency of the synthetic gain curve, there is no case where the original hole of one frequency burns overlaps with the burned image hole of the other frequency on the same gain curve (Zhang Shulian. 2005). Therefore, the influence of the $(\omega - \Delta\upsilon_1)$ term on the mutual saturation coefficient can be ignored, and the magnitude of the split frequency difference mainly determines the magnitude of θ.

The optical length of the S sub-cavity is about 104.8 mm, the longitudinal mode spacing is 1431 MHz; the optical length of the P sub-cavity is about 102.9 mm, and the longitudinal mode spacing is 1457 MHz. Although the loss of the longitudinal mode in the S and P cavities is different, the output bandwidth is basically about 1500 MHz, which is slightly larger than longitudinal mode spacing. At the same time, the frequency difference blocking threshold measured in the experiment is about 33 MHz, so there are at most three longitudinal modes in the output bandwidth, two of which are longitudinal modes of the same polarization.

To summarize, the light intensity tuning curve is mainly influenced by four factors:

1) Linear net gain coefficient α: The closer to the longitudinal mode of the gain center frequency, the larger the linear net gain coefficient α is.
2) Self-saturation coefficient β: The closer to the longitudinal mode of the gain center frequency, the larger β is.
3) Single-pass loss factor γ: As the piezoelectric ceramic voltage increases, the displacement variation of PZT changes nonlinearly, resulting in a nonlinear change in the single-path loss factor of each longitudinal mode.
4) Mutual saturation coefficient θ: In the public cavity tuning process, the mutual longitudinal saturation coefficient θ of adjacent longitudinal modes in the case of different split frequency differences is different, which leads to the competition strength and competition result are different. If the split frequency difference is larger, the mutual saturation coefficient θ will be smaller, and the light intensity will be stronger.

Therefore, in the public cavity tuning process, the linear gain and self-saturation effect, loss and mutual saturation effect of each longitudinal mode are combined to affect the change of light intensity in the self-consistent equation of light intensity.

4 EXPERIMENTAL RESULTS AND ANALYSIS

4.1 *Light intensity tuning curve and analysis*

Different frequency differences are obtained by changing the voltage of PZT2 in Figure 1. In the case of different frequency differences, different light intensity tuning curves are obtained by changing the voltage of the piezoelectric ceramic PZT1 on the public cavity. The experimental results show that the frequency difference blocking threshold is around 33 MHz when the electric current is 1.8 mA. In the case of different split frequency differences, three sets of intensity tuning curves were selected and analyzed from the experimental data. The split frequency differences are as follows: $\Delta v < 33MHz$, $\Delta v = 86MHz$, $\Delta v = 665MHz$.

4.1.1 *The intensity tuning curve and the beat frequency when frequency difference is blocked*

Figure 2a is a graph showing the change of light intensity of Slight and P light when the frequency difference is blocked. The abscissa is the voltage of the piezoelectric ceramic PZT1 on the public cavity, and the ordinate is the light intensity (the same in the next two sections). At the same time, the beat frequencies are shown in Figure 2b, the abscissa is the voltage of PZT1, and the ordinate is the beat frequency (the same in the next two sections).

According to Figure 2, experimental phenomena and theoretical analysis are as follows:

1) The light intensity variation curve of S light is approximately sinusoidal, and the light intensity of P light is substantially zero. When the split frequency difference is less than the blocking threshold, the mode competition is a strong competition. It is assumed that the intensity of S light and P light are equal at the initial time, the gain coefficients of the two are approximately equal because of the small frequency difference. The loss of P light is larger than that of S light, so the light intensity increase rate of S light is greater than that of P light. Because the mutual saturation coefficient θ is large, the P light is finally rapidly reduced to extinction. The final result of the mode competition is that the S light is oscillating and the P light is extinguished.

2) In the experiment, only one intermittent beat frequency whose value is 1431 MHz appeared, which is the longitudinal mode spacing of S light. At this time, the double longitudinal mode of S light oscillates at both edges of the gain curve, while P light is extinguished. As the voltage of PZT1 increases, the "S mode" on the left side is closer to the center frequency of the gain curve, while the adjacent "S mode" is outside the output bandwidth, causing the beat frequency to disappear. In the stage of no beat frequency, only the S light single longitudinal mode oscillates at this time, while P light is extinguished.

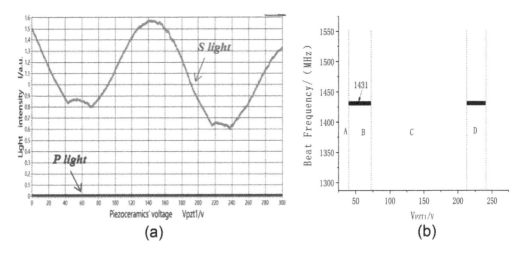

(a)   (b)

Figure 2. Experimental tuning curves of light intensity (a) and beat frequencies diagram (b) when the frequency difference is blocked.

4.1.2 *The intensity tuning curve and the beat frequency when frequency difference is 86 MHz*
According to Figure 3, the distribution of the longitudinal modes on the gain curve of each beat frequency stage can be obtained, as shown in Figure 4. Experimental phenomena and theoretical analysis are as follows:

1) According to Figure 4, from stage A to stage B, as the voltage of PZT1 increases, the gain coefficient of the "P mode" decreases continuously until the loss is greater than the gain obtained, resulting in P light being extinguished in stage B, while the intensity of S light gradually decreases as the voltage of the PZT1 increases. At this time, only single "S mode" oscillates in stage B, and there is no beat frequency.

2) In stage C, "P-1 mode" starts to oscillate. As time goes by, the gain coefficient of "P-1 mode" increases while that of "S mode" decreases continuously, resulting in a rapid decrease of the intensity of S light while intensity of P light is increasing rapidly. Only the "S mode" and "P-1 mode" oscillate at the same time, and the beat frequency is 1371 MHz.

3) In stage D, "S-1 mode" enters the output bandwidth and begins to oscillate. Although the gain coefficient of "P-1 mode" is always larger than that of "S-1 mode", the loss of "P-1 mode" is much larger than "S-1 mode". This causes "S-1 mode" to "grab" more gain particles than "P-1 mode", so the increase of intensity of P light is not obvious.

Three different beat frequencies appear, which are 1431 MHz, 1371 MHz, and 60 MHz, respectively. The analysis of other stages is the same as above.

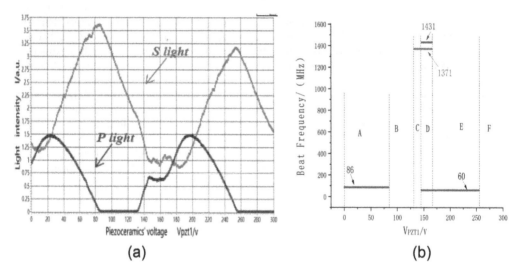

Figure 3.   Experimental tuning curves of light intensity (a) and beat frequencies diagram (b) when the frequency difference is 86 MHz.

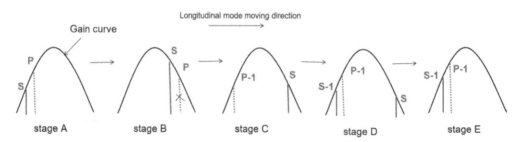

Figure 4.   Longitudinal mode distribution diagram of each stage in Figure 3b.

80

4.1.3 *The intensity tuning curve and the beat frequency when frequency difference is 665 MHz*

According to the relationship between longitudinal mode spacing and output bandwidth, the distribution of the longitudinal modes on the gain curve of each beat frequency stage can be obtained, as shown in Figure 6. Experimental phenomena and theoretical analysis are as follows:

In stage A, as the voltage of PZT1 increases, the light intensity of the S light gradually increases while the light intensity of the P light decreases continuously until the P light is extinguished in stage B. The change of light intensity in stage C is just the opposite in stage B. As shown in Figure 6, in stage A, "P+1 mode" and "S+2 mode" oscillate, whose frequency difference is 665 MHz. In stage B, only "P+1 mode" oscillates while "S+2 mode" is outside the output bandwidth. That is the reason why there is no beat frequency in stage B. The analysis of other stages is the same as above.

Summary of the basic laws of light intensity tuning:

1) Regardless of the frequency difference, the light intensity tuning curve of P light and S light are approximately sinusoidal, and the change trend is reversed.
2) The light intensity tuning curve of P light and S light are crossed, when the frequency difference is blocked is not considered.
3) During the tuning of the public cavity, there is always a state in which a certain polarized light is extinguished.
4) The size of the split frequency difference is the main factor affecting the change of the light intensity tuning curve.
5) After many experiments, when the split frequency difference is in the range of 129—1302 MHz, the intensity tuning curve is similar to that in Figure 5a. There is only one S longitudinal mode and one neighboring P longitudinal mode in the output bandwidth.

4.2 *Frequency difference tuning curve and analysis*

By tuning the public cavity or S sub-cavity, the double S longitudinal modes are stabilized at both edges of the gain curve, and then the P sub-cavity is continuously tuned to obtain a frequency difference tuning curve, as shown in Figure 7. Figure 8 is an enlarged view of the black dotted frame in Figure 7. In both figures, the abscissa is the voltage of piezoelectric ceramic on the P sub-cavity, the ordinate is the beat frequency, and the four curves of a, b, c, and d are the sweep curves. The longitudinal mode distribution of the tuning process is shown in Figure 9, where the "S mode" and "S+1 mode" are fixed at the point v_s and v_{s+1} in the gain curve, respectively.

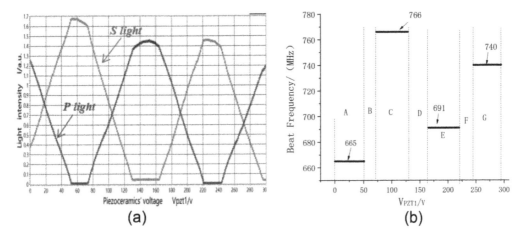

Figure 5.   Experimental tuning curves of light intensity (a) and beat frequencies diagram (b) when the frequency difference is 665 MHz.

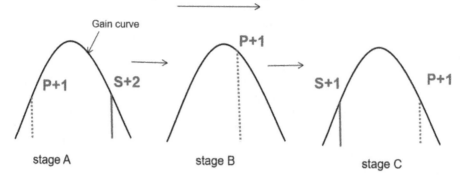

Figure 6. Longitudinal mode distribution diagram of each stage in Figure 5b.

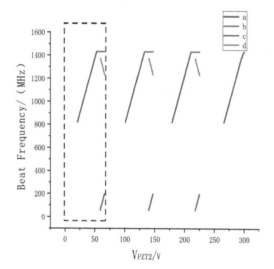

Figure 7. Frequency difference tuning curve.

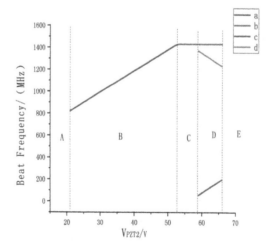

Figure 8. Frequency difference tuning curve in dotted frame in Figure 7.

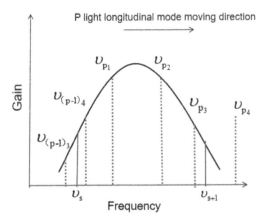

Figure 9. Longitudinal mode distribution diagram.

Experimental phenomena and theoretical analysis are as follows:

1) There is no beat frequency in stage A: only a "P mode" oscillates in the position of υ_{p1} to υ_{p2} of the gain curve in Figure 9. When V_{PZT2} changes from 0V, "P mode" moves from the point of υ_{p1} to the right, along the gain curve. In stage A, "P mode" is always close to the center frequency of the gain curve. Although the loss of P light in the cavity is greater than that of the S light, the gain of the S light is far less than that of the P light since the double S longitudinal modes are at both edges of the gain curve. So the double S longitudinal modes cannot oscillate, while only a "P mode" oscillates.

2) A beat frequency linearly increases from 820 MHz to 1431 MHz in stage B: "P mode" is in the position of υ_{p2} to υ_{p3}. When "P mode" passes through the center of the gain curve, the gain coefficient of "P mode" decreases continuously, and the split frequency difference between
"S mode" and "P mode" becomes larger, which makes the mutual saturation coefficient θ become smaller. Therefore, the number of gain particles obtained by "P mode" on the left side is continuously increased until "P mode" passes through the point υ_{p2}, the gain obtained by "S mode" is greater than the loss, and S light starts to oscillate. The competition between "S+1 mode" and "P mode" is more intense and "S+1 mode" still cannot oscillate. The beat frequency we observe is generated by "P mode" and "S mode".

3) The analysis of other stages is the same as above. In stage C, "P mode" is in the position of υ_{p3} to υ_{p4}, while "P-1 mode" is in the position of $\upsilon_{(p-1)3}$ to $\upsilon_{(p-1)4}$. But both of them cannot oscillate because of the failure in the competition between "S+1 mode" and "S mode", respectively. The beat frequency we observe is generated by "S+1 mode" and "S mode". In stage D, beat frequency b is generated by "S+1 mode" and "S mode", beat frequency c is generated by "P-1 mode" and "S+1 mode", and beat frequency d is generated by "P-1 mode" and "S mode".

5 CONCLUSION

In this paper, the intensity tuning curve, the corresponding beat frequency curve and the frequency difference tuning curve of the new integrated lasers with Y-shaped cavity in the case of different split frequency differences are reported. The theoretical analysis and interpretation are carried out as well. By using the Lamb semi-classical gas laser theory within the third-order perturbation approximation, the influencing factors and mechanism of the mode competition and cavity tuning characteristics are analyzed and summarized comprehensively. The analysis believes that split frequency difference is the main factor affecting the change of the light intensity tuning curve. Affected by the split frequency difference, in the public cavity tuning process, the linear gain and self-saturation effect, loss and mutual saturation effect of each longitudinal mode are combined to affect the change of light intensity in the self-consistent equation of light intensity. The influencing factors and tuning laws of the light intensity tuning curve are summarized. In the end, the experiment verifies that when the split frequency difference is in the range of 129—1302 MHz, the laser is basically in the working state which the single longitudinal mode pair is oscillating. The experimental results and theory lay the foundation for the research in the stability improvement of the beat frequency of the new laser in the next work.

REFERENCES

Gong Mengfan. 2015. "Investigation on the Physical Characteristics of Orthogonal Polarized He-Ne Laser with Integrated Y-shaped Cavity", National University of Defense Technology.

Guangzong Xiao. 2011. "Preliminary study on laser accelerometer based on Y-shaped cavity orthogonal polarized dual-frequency laser", National University of Defense Technology.

Han Yanmei, Zhang Shulian, Li Kelan. 1997. "Power tuning for 6328 nm wavelength He-Ne lasers with various frequency spacing by mode split", Laser Technology, 21(2): 111~114.

Han Yanmei, Zhang Shulian, Li Yan et al. 1998. "Theoretical analysis of the laser power tuning property", Laser Technology, 22(4): 211~214.

Jin Yuye, Zhang Shulian, Li Yan et al. 2001. "Zeeman birefringence dual frequency laser", Chin. Phys. Lett., 18(4): 533~536.

Lamb W.E. 1964. "Theory of an optical maser", Physical Review 1, 34(6): A14290~A1450.

Lu Yaxiong, Yu Xuecai, Zhang Xiaoxia. 2005. "Laser Physics", Beijing University of Posts and Telecommunications Press.

T. Yoshino, M. Kawata, B. Qimude et al. 1998. "Fiber-coupling-operated orthogonal-linear-polarization Nd: YAG micro chiplaser: photothermal beat-frequency stabilization and interferometric displacement measurement application", J. Lightwave Technol., 16(3): 453~458.

T. Yoshino. 2008. "Performance analysis of intracavity birefringence sensing", Appl. Opt., 47(14): 2655~2659.

Xiao Guangzong, Long Xingwu, Zhang Bin, Lu Guangfeng, Zhao Hongchang. 2011. "Mode Competition and Laser Power Tuning Property of Orthogonal Polarized Laser with Y-Shaped Cavity", Chinese Journal of Lasers, 38(03): 85–90.

Yang Sen, Zhang Shulian. 1988. "The frequency split phenomenon in a He-Ne laser with rotation quartz crystal plate in its cavity". Opt. Commun, 68(1): 55~57.

Zhang Shulian, Lu Men, Wu Minxian et al. 1993. "Laser frequency split by an electron-optical element in its cavity". Opt. Commun., 96(4): 245~248.

Zhang Shulian. 2005. "The Theory of Orthogonal Polarized Laser", Tsinghua University Press.

Advances in Optoelectronic Technology and Industry Development – Jose & Ferreira (eds)
© *2020 Taylor & Francis Group, London, ISBN 978-0-367-24634-1*

Effects of pressure on the femtosecond filamentation with HOKE in air

X.X. Qi & C.R. Jing
College of Physical & Electronic Information, Luoyang Normal University, Luoyang, China

ABSTRACT: We investigate the pressure effects on the propagation of the intense femtosecond laser pulse with wavelength of 800 nm by numerical simulations. We adopt the higher-order Kerr model and consider the effects on the on-axis intensity, the beam radius and the energy of the filament, as well as the on-axis electron density. Numerical results show that when the pressures increase, the filament appears later and ends earlier resulting in the shorter filament length. The cross-sectional radius of the filament becomes narrower with the increase of pressure. We also obtain the conclusion that the energy in the filament background energy pool increases when the pressure increases.

1 INTRODUCTION

Since Braun et al. carried out the experiment with an intense infrared (IR) femtosecond laser pulse demonstrating that the intense ultra-short laser pulse is suited for long range propagation in air (Braun et al. 1995), a large amount of research had been engaged in this propagating type which is called filamentation or self-guided propagation. The physical mechanism of filamentation is interpreted as a dynamic equilibrium between Kerr self-focusing and plasma defocusing in classical model (Chekalin et al. 2019, Li et al. 2014, Kohler et al. 2013, Kosareva et al. 2011, Couairon & Berge 2000) or between Kerr self-focusing and defocusing by higher-order Kerr effect (HOKE) in higher-order Kerr model (Qi et al. 2019, Qi et al. 2016, Milchberg 2014, Petrarca et al. 2012, Bejot et al. 2011, Carsten et al. 2011, Béjot et al. 2010).

The filamentation at 1 atm (standard atmosphere pressure) has been studied intensively in both the classical model and the higher-order Kerr model. For instance, based on the classical model, Xi et al. discover that fusing two in-phase light filaments can form a long and stable channel (Xi et al. 2006), and in the stable channel, multiphoton ionization (MPI) plays an essential role (Chiron et al. 1999). The path length of the light bullets amounts to about 200 mm and does not depend on the pulse energy and the way of focusing (Chekalin et al. 2019). Béjot et al. obtain that two-color co-filament can provide a potential way to generate an attosecond pulse (Béjot et al. 2008). For the full model, the HOKE can arrest the self-focusing collapse at a lower intensity (Kosareva et al. 2011) and as the competitive counterpart of focusing, it significantly affects the first self-focusing stage to form a ring-shaped pattern (Wang et al. 2010, Wang et al. 2011). The incident pulse duration has a significant influence on the relative contribution of HOKE (Jing et al. 2019).The filament can be re-produced better with the full model than the classical model (Petrarca et al. 2012). In filamentation dynamics, HOKE can act similar to plasma under certain predesigned conditions (Wang et al. 2013).

The filamentation under the non-standard-atmosphere pressures has also been studied numerically based on the classical model. When the pressure is about 10 atm, the GVD has a great influence on the collapse distance (Li et al. 2014). The laser noise is compressed by filamentation in argon at about 5 atm pressure (Béjot et al. 2007). The pulse with the duration as short as 2.5 attoseconds can be achieved under 80 atm (Popmintchev et al. 2012). The influence of the input beam shape, chirp and pulse duration on the length of the filament channels is discussed from 0.2 atm to 1 atm (Couairon et al. 2006), and the filament length becomes more homogeneous at higher altitude when the input peak power is fixed (Hosseini et al. 2012).

In this work, we study the filamentation at different pressures including the HOKE. The content is organized as follows. Section 2 introduces the nonlinear Schrödinger equation for the femtosecond pulse propagation in air at different pressures and the computational algorithm employed in the simulations. In Section 3, the effects of pressures on the filament in air are investigated. The conclusions are summarized in Section 4.

2 MODEL AND ALGORITHM

Expressed in the reference frame moving at the group velocity, the propagation equation of a laser pulse with a linearly polarized incident electric filed $E(r, t, z)$ along the propagation direction z reads (Jing et al. 2019, Wang et al. 2010, Couairon et al. 2010, Xi et al. 2006, Couairon et al. 2002):

$$\frac{\partial E}{\partial z} = \frac{i}{2k_0} \left(\frac{\partial^2}{\partial r^2} + \frac{1}{r} \frac{\partial}{\partial r} \right) E - \frac{ik''}{2} \frac{\partial^2 E}{\partial t^2} + \frac{ik_0}{n_0} \Delta n_{kerr} E - \frac{ik_0}{2} \frac{\omega^2}{\omega_0^2} E - \frac{\beta^{(k)}}{2} |E|^{2K-2} E \quad (1)$$

where t refers to the retarded time in the reference frame of the pulse ($t \rightarrow t - \frac{z}{v_g}$ with $v_g = \frac{\partial \omega}{\partial k} |\omega_0$ corresponding to the group velocity of the carrier envelope). The first two terms on the right-hand side of Equation (1) are the linear effects, accounting for the spatial diffraction and the second order dispersion. The other three terms represent Kerr effect, plasma defocusing effect and the multiphoton ionization effect, and these terms are the nonlinear effects. In Equation (1), $k_0 = 2\pi/\lambda$ and $\omega_0 = 2\pi c/\lambda$ are the wave number and the angular frequency of the carrier wave, respectively. k'' is the second order dispersion coefficient. The nonlinear index Δn_{kerr} induced by intense femtosecond laser pulses can be written as $\Delta n_{kerr} = n_2 |E|^2 + n_4 |E|^4 + n_6 |E|^6 + n_8 |E|^8 + ...$, where n_{2*j} are coefficients that have been reported in references (Loriot et al. 2009, Loriot et al. 2010) are related to $\chi^{(2*j+1)}$ susceptibilities. The plasma frequency is $\omega = \sqrt{\rho e^2/m_e \varepsilon_0}$ with e, m_e, ρ being the electron charge, mass and density. $\beta^{(K)}$ denotes the nonlinear coefficient for K-photon absorption, where K is the minimal number of photons needed to ionize gas.

The evolution of electron density ρ follows the equation (Couairon et al. 2010, Bejot et al. 2007):

$$\frac{\partial \rho}{\partial t} = \frac{\beta^{(K)} |E|^{2K}}{K \hbar \omega_0} - \frac{\beta^{(K)} |E|^{2K}}{K \hbar \omega_0} \frac{\rho}{\rho_{at}} \quad (2)$$

in which the right-hand terms consist of the photo-ionization, the avalanche ionization, the electron attachment and the electron-ion recombination. In Equation (2), U_i is the ionization potential, $\hbar = h/2\pi$ with h the Planck constant, the cross-section σ for inverse bremsstrahlung at 1 atm (standard atmosphere pressure) follows the Drude model (Yablonovitch et al. 1972) and $\sigma = \frac{k_0 e^2}{\omega_0^2 \varepsilon_0^2} \frac{\omega_0 \tau_c}{\omega_0^2 \tau_c^2}$ ($\tau_c = 350$ fs is the relaxation time) (Couairon et al. 2006), ρ_{at} denotes the neutral particle density of gas under 1 atm.

The input electric field envelop is modeled by a Gaussian profile with input power P_{in} as

$$E(r, z, t)|_{z=0} = \sqrt{\frac{2P_{in}}{\pi r_0^2}} \exp\left(-\frac{t^2}{t_0^2}\right) \exp\left(-\frac{r^2}{r_0^2}\right) \quad (3)$$

where r_0 and t_0 denote the radius and duration of the pulse, respectively.

The model is solved via the Split-Step Fourier method (Agrawal 2006), in which all the linear terms are calculated in the Fourier space over a half-step while the nonlinear terms are calculated in the physical space over a second half-step. To integrate the linear part of the Eq. (1) along the propagation axis, we adopt the Crank-Nicholson scheme (Press et al. 2007), which is more stable than the Euler method (Chiron et al. 1999). For Equation (2), the 4th-order Runge-Kutta method is employed.

3 RESULTS AND DISCUSSION

We consider an incident laser pulse with the wavelength $\lambda = 800$ nm, the duration $t_0 = 100$ fs and the beam radius $r_0 = 1$ mm. The power of the incident laser pulse is set as $P_{in} = 3P_{cr}$, where $P_{cr} = 3.77\lambda^2/8\pi n_2$ is the critical power.

In this paper the beam radius of the incident pulse is defined as e^{-2} of maximum intensity of the pulse (Hao et al. 2009), so is the radius of the filament in the propagation of pulse. Figure 1 presents the evolution of laser on-axis intensity, beam radius, the ratio of energy of filament to energy of the total area, as well as the on-axis plasma density at different pressures.

From Figure 1a, it can be seen that when the pressure changes from 0.5 atm to 4 atm, although the positions of the maximal on-axis intensity are different - the position is the latest at 4 atm and the position of 0.5 atm is the earliest, the maximal values are stable around 3×10^7 W/m^2, which is consistent with the light intensity clamping experiment carried out by Bernhardt

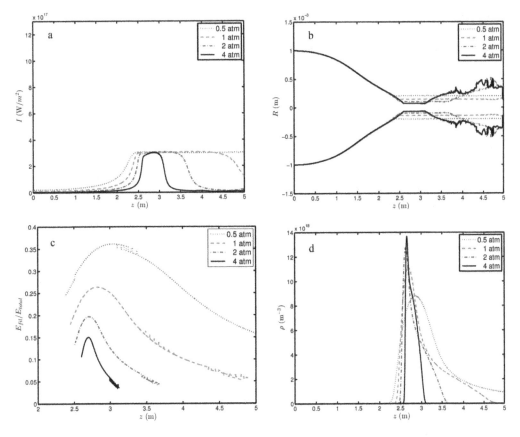

Figure 1. Evolution of the pulse and the on-axis plasma density with the propagation distance z in the classical model. (a) The on-axis intensity of pulse; (b) beam radius; (c) the ratio of energy of filament to the total energy; and (d) the on-axis plasma density.

et al. (Bernhardt et al. 2008). In reality, when intense femtosecond laser propagates vertically in the air, the diameter of the filament increases with the increase of altitude, because the intensity clamping is independent of the pressure. The self-focusing critical power is inversely proportional to the non-linear Kerr refractive index, and the non-linear refractive index is positively related to the air density, so the self-focusing critical power will increase at high altitude. Therefore, in order to achieve self-focusing at high altitude, the input peak power of laser pulse should be increased. However, the phenomenon of light intensity clamping has occurred at sea level, so laser pulses at high altitude need larger transverse dimensions to obtain higher peak power relative to the standard atmospheric pressure at sea level. From Figure 1a, it can be obtained that the propagation distances of peak light intensity are different under four atmospheric pressures. The propagation distance of 0.5 atm is the longest and that of 4 atm is the shortest, which is also reflected in Figure 1b.

Figure 1b shows the variation of beam radius with propagation distance simulated by the higher-order Kerr model. It can be seen that under the four pressure environments, the intense femtosecond laser starts focusing from the initial radius of 1 mm, and the focusing trajectories are the same, but the starting and ending positions of the filament are different. The higher the pressure is, the later the filament appears and the earlier the filament ends, resulting in the shortest filament length. The cross-sectional radius of the filament becomes narrower and narrower with the increase of pressure, that is, the length and radius of the filament both decrease with the increase of pressure.

The variation of the ratio of the energy in the filament to the total energy of the beam with the propagation distance is shown in Figure 1c. The results of standard atmospheric pressure conditions are in great agreement with those in references (Liu et al. 2005, Hao et al. 2009). It can be seen that in the filament length range, the maximal ratios of the energy in the filament to the total energy of the beam are: 0.5 atm, 35%; 1 atm, 26%; 2 atm, 19% and 4 atm, 15%. The ratio of the energy in the filament to the total energy of the beam increases first and then decreases along the propagation direction, and the ratio decreases with the increase of the pressure, which means that the energy in the filament background energy pool increases with the increase of the pressure.

The variation of electron density with propagation distance simulated in four pressure environments is shown in Figure 1d. It can be seen that the electron density of the black solid line at 4 atm is the highest, approximately $13.7 \times 10^{18}/m^3$, with the decrease of pressure, the electron density decreases, and highest electron density of 0.5 atm is about $8.6 \times 10^{18}/m^3$. It can also be seen from the figure that the electron density at 4 atm decreases most quickly from the highest value, the curve slope is steepest, and the electron density at 0.5 atm decreases slowly from the highest value, especially after 3.7 m, it decreases more slowly.

4 CONCLUSIONS

We have investigated the effects of pressure on the filamentation numerically based on the higher-order Kerr model. The results show that the pressure plays an important role in the formation of femtosecond filaments. When the pressures increase, the length and the cross-sectional radius of the filament become shorter and narrower, respectively. The maximal electron density gets higher with the increasing pressure we consider. We also obtain that the energy in the filament decreases implying the energy in the background energy pool increases with the increase of the pressure.

ACKNOWLEDGMENTS

The authors thank the reviewer for her/his constructive comments and suggestions on improving the quality of this paper. This work is supported by the National Natural Science Foundation of China (Grant No. 11704174).

REFERENCES

Agrawal, G.P. 2006. *Nonlinear fiber optics.* Salt Lake City: Academic Press.

Béjot, P., Bonnet, C., Boutou, V. & Wolf, J.P. 2007. Laser noise compression by filamentation at 400 nm in argon. Optics Express, 15(20), 13295–13309.

Béjot, P., Kasparian, J. & Wolf, J.P. 2008. Dual-color co-filamentation in Argon. Optics Express, 16(18): 14115–14127.

Béjot, P., Kasparian, J., Henin, S., Loriot, V., Vieillard, T. & Hertz, E. 2010. Higher-order kerr terms allow ionization-free filamentation in gases. Physical Review Letters, 104(10), 103903.

Béjot, P., Hertz, E., Kasparian, J., Lavorel, B., Wolf, J.P. & Faucher, O. 2011. Transition from plasma-driven to kerr-driven laser filamentation. Physical Review Letters, 106(24), 243902.

Bernhardt, J., Liu, W., Chin, S.L., & Sauerbrey, R. 2008. Pressure independence of intensity clamping during filamentation: theory and experiment. Applied Physics B, 91(1), 45–48.

Braun, A., Korn, G., Liu, X., Du, D., Squier, J. & Mourou, G., 1995. Self-channeling of high-peak-power femtosecond laser pulses in air. Optics Letter, 20 (1), 73–75.

Carsten B., Ayhan D. & Günter S. 2011. Saturation of the All-Optical Kerr Effect. Physical Review Letters, 106(18), 183902.

Chekalin, S.V., Kompanets, V.O., Zaloznaya, E.D. & Kandidov V.P. 2019. Effect of group velocity dispersion on femtosecond filamentation of Bessel–Gaussian beams. 49(4), 344–349.

Chiron, A., Lamouroux, B., Lange, R., Ripoche, J.F., Franco, M. & Prade, B., et al. 1999. Numerical simulations of the nonlinear propagation of femtosecond optical pulses in gases. The European Physical Journal D - Atomic, Molecular, Optical and Plasma Physics, 6(3), 383–396.

Couairon, A. & Berge, L., 2000. Modeling the filamentation of ultra-short pulses in ionizing media. Phys. Plasmas 7 (1), 193–209.

Couairon, A. & Bergé, L. 2002. Light filaments in air for ultraviolet and infrared wavelengths. Physical Review Letters, 88(13), 135003.

Couairon, A., Franco, M., Méchain, G., Olivier, T., Prade, B. & Mysyrowicz, A. 2006. Femtosecond filamentation in air at low pressures: part i: theory and numerical simulations. Optics Communications, 259(1), 265–273.

Couairon, A. & Mysyrowicz, A. 2010. Femtosecond filamentation in transparent media. Physics Reports, 441(2), 47–189.

Köhler, C., Guichard, R., Lorin, E., Chelkowski, S., Bandrauk, A. D. & Bergé, L., et al. 2013. On the saturation of the nonlinear refractive index in atomic gases. Physical Review A, 87(4), 043811.

Kosareva, O., Daigle, J.F., Panov, N., Wang, T., Hosseini, S. & Yuan, S., et al. 2011. Arrest of self-focusing collapse in femtosecond air filaments: higher order Kerr or plasma defocusing? Optics Letters, 36(7), 1035–1037.

Hao, Z., Zhang, J., Lu, X., Xi, T., Zhang, Z. & Wang, Z. 2009. Energy interchange between large-scale free propagating filaments and its background reservoir. Journal of the Optical Society of America B, 26(3), 499–502.

Hosseini, S., Kosareva, O., Panov, N., Kandidov, V.P., Azarm, A. & Daigle, J.F., et al. 2012. Femtosecond laser filament in different air pressures simulating vertical propagation up to 10 km. Laser Physics Letters, 9(12), 868–874.

Jing, C.R., Qi, X.X., Wang, Zh. H., Ma, B.H. & Ding, Ch. L. 2019. Comparative study of femtosecond filamentation properties in the classical model and the full model for different incident pulse durations. Journal of Optics, 21, 065503.

Li, S.Y., Guo, F.M., Song, Y., Chen, A.M., Yang, Y. J. & Jin, M.X. 2014. Influence of group-velocity-dispersion effects on the propagation of femtosecond laser pulses in air at different pressures. Physical Review A, 89(89), 023809.

Liu, W., F. Théberge, E. Arévalo, Gravel, J.F., Becker, A. & Chin, S.L. 2005. Experiment and simulations on the energy reservoir effect in femtosecond light filaments. Optics Letters, 30(19), 2602–2604.

Loriot, V., Hertz, E., Faucher, O. & Lavorel, B. 2009. Measurement of high order Kerr refractive index of major air components. Optics Express, 17(16), 13429–13434.

Loriot, V., Hertz, E., Faucher, O. & Lavorel, B. 2010. Measurement of high order Kerr refractive index of major air components. Optics Express, 18(3), 3011–3012.

Milchberg, H. 2014. The extreme nonlinear optics of gases and femtosecond optical/plasma filamentation. Physics of Plasmas, 21(10), 47–189.

Petrarca, M., Petit, Y., Henin, S., Delagrange, R., Béjot, P. & Kasparian, J. 2012. Higher-order kerr improve quantitative modeling of laser filamentation. Optics Letters, 37(20), 4347–4349.

Popmintchev, T., Chen, M.C., Popmintchev, D., Arpin, P., Brown, S. & Alisauskas, S., et al. 2012. Bright coherent ultrahigh harmonics in the kev x-ray regime from mid-infrared femtosecond lasers. Science, 336(6086), 1287–1291.

Press, W.H., Flannery, B.P., Teukolsky, S.A. & Vetterling, W.T. 2007. *Numerical recipes*. New York: Cambridge University Press.

Qi, X., Ma, C. & Lin, W. 2016. Pressure effects on the femtosecond laser filamentation. Optics Communications, 358, 126–131.

Qi, X. & Jing Ch. R. 2019. Effects of higher-order Kerr on the femtosecond filament in argon. Laser and Infrared, 49(1), 31–34.

Wang, H., Fan, C., Zhang, P., Qiao, C., Zhang, J. & Ma, H. 2010. Light filaments with higher-order Kerr effect. Optics Express, 18(23), 24301–24306.

Wang, H., Fan, C., Shen, H., Zhang, P.F., Qiao, C.H. & Ma, H. 2011. Dynamics of femtosecond filamentation with higher-order Kerr response. Journal of the Optical Society of America B, 28(9), 2081.

Wang, H., Fan, C., Shen, H., Zhang, P.F. & Qiao, C.H. 2013. Relative contributions of higher-order Kerr effect and plasma in laser filamentation. Optics Commmunications, 293, 113–115.

Xi, T.T., Lu, X. & Zhang, J. 2006. Interaction of light filaments generated by femtosecond laser pulses in air. Physical Review Letters, 96(2), 025003.

Yablonovitch, E. & Bloembergen, N. 1972. Avalanche ionization and the limiting diameter of filaments induced by light pulses in transparent media. Physical Review Letters, 29(14), 907–910.

Optical communications

Advances in Optoelectronic Technology and Industry Development – Jose & Ferreira (eds)
© 2020 Taylor & Francis Group, London, ISBN 978-0-367-24634-1

Performance investigation of 16/32-channel DWDM PON and long-reach PON systems using an ASE noise source

D. Shanmuga Sundar
Universidad de Chile, Santiago, Chile

T. Sridarshini
Madras Institute of Technology, Chennai, Tamilnadu, India

R. Sitharthan
Madanapalle Institute of Technology and Science, Madanapalle, Andhra Pradesh, India

Madurakavi Karthikeyan
Vellore Institute of Technology, Vellore, Tamilnadu, India

A. Sivanantha Raja
Alagappa Chettiar Government College of Engineering and Technology, Karaikudi, Tamilnadu, India

Marcos Flores Carrasco
Universidad de Chile, Santiago, Chile

ABSTRACT: One of the major worldwide problems is a growing demand for transmission capacity. One solution can be provided by the use of spectrum-sliced Dense Wavelength-Division Multiplexed (DWDM) optical networks, for their cost-effective and proficient power solution. In this paper, we successfully designed and demonstrated 16/32-channel DWDM Passive Optical Network (PON) and long-reach PON systems for 2.5 and 10 Gbps bit rates using Amplified Spontaneous Emission (ASE) noise generated by an erbium-doped fiber amplifier. In addition, we have analyzed the best modulation format for a 10 Gbps/32-channel DWDM long-reach PON system. The performance of the intended ASE-based DWDM system is scrutinized in terms of eye diagrams, Bit Error Rates (BERs), Q-values, and so on. An improved BER is achieved.

1 INTRODUCTION

The demand for broadband access is due to the rapid growth of Internet access, such as IP video delivery and Voice-over-IP (VoIP). In order to provide a solution for these problems, broadband services provided by using copper access networks are being replaced by the optical access technology that has been commercially available for several years and is being deployed in some countries (Lin, 2006). Wavelength-Division Multiplexed Passive Optical Networks (WDM-PONs) are considered to be a futureproof solution for PONs because of their ideal features, such as upgradability, high network capacity and flexibility (Lee et al., 2012). PONs are often the technology of choice because the transmission fiber and the central office equipment can be shared by a large number of customers when deployed (Cauvin et al., 2006).

Traditional WDM systems have multiple transmitter lasers operating at different wavelengths, which need to be wavelength-selected for each individual channel operated at a specific wavelength (El-Sahn et al., 2010; Lee et al., 2012). This increases the complexity of network architecture, cost and wavelength management (Choi & Lee, 2011). The spectrum-sliced Dense Wavelength-Division Multiplexed passive optical network (DWDM-PON) is a cost-effective and power-efficient solution for optical access networks in satisfying the growing worldwide demand for transmission capacity (Kaneko et al., 2006; Spolitis & Ivanovs, 2011). Spectrum-sliced optical systems benefit from the same advantages as WDM, while employing low-cost incoherent broadband light sources such as broadband-Amplified Spontaneous Emission (ASE) sources or Light-Emitting Diodes (LEDs) (Kaneko et al., 2006). The method of spectrum slicing is a cost-efficient, appropriate and promising solution for transmitters in the Optical Line Terminal (OLT) of DWDM-PON systems (El-Sahn et al., 2010).

The concept of increasing the reach and/or split of PONs via intermediate equipment such as optical amplifiers has been of research interest since the 1990s (Forrester et al., 1991). Recently, research has focused on extending the reach of Gigabit PONs (G-PONs) and Gigabit Ethernet PONs (GE-PONs) via mid-span optical amplifiers (Suzuki et al., 2007) or transponders (Davey et al., 2006). This concept has recently been standardized in ITU-T Recommendation G.984.6. The OLT is connected via a length of fiber known as the Optical Trunk Line (OTL) to the active mid-span extender equipment. This in turn is connected to the Optical Distribution Network (ODN) and Optical Network Unit (ONU). In this paper, we demonstrate a new model of 16/32-channel 2.5/10 Gbps DWDM-PON and long-reach PON using single broadband ASE with a flat spectrum. In general, a high-power ASE source with flat spectrum is most suitable for use in homogeneous WDM optical systems because it is very important to obtain spectrally sliced channels with equivalent output power levels. In addition, we evaluated the performance of the designed systems with different modulation formats.

2   GENERATION OF ASE SOURCE

In this system, ASE noise generated by an Erbium-Doped Fiber Amplifier (EDFA) using a light source with pump signals has been used as a Broadband Light Source (BLS). The signals obtained are of different powers; in order to avoid power channel separation and power fluctuations, the signals are passed through a gain-flattening filter and are optimized to obtain a flat spectrum for better performance. In this system, a flat spectrum of power level -20 dB is achieved. The flat spectrum can also be obtained by using the EDFA in a cascaded configuration but better results can be achieved by using an optimized pump power EDFA with gain-flattening filter.

Through utilization of our reported configuration, the cost of the WDM system can be reduced significantly. Figures 1a and 1b represent the generated and flattened ASE signal

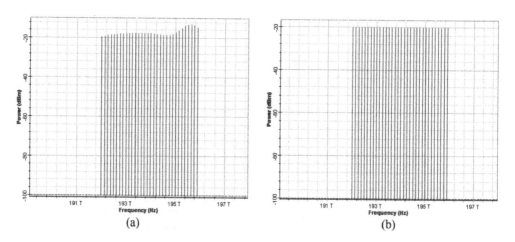

Figure 1.   ASE signal source in: (a) generated form; (b) flattened form.

source (i.e. the spectrum before and after the gain-flattening filter). The generated flattened ASE broadband source is then sliced into individual channels by using one of the basic spectrum-slicing techniques, which are known for their cost-effectiveness and simplicity. The main aim of the spectrum slicing is to deploy a single incoherent broadband source into a large number of wavelength channels. This technique is also reported as a good candidate for generating equally spaced multi-wavelength channels (Lee et al., 2012).

3 16- AND 32-CHANNEL (2.5/10 GBPS) WDM PON AND LONG-REACH PON SYSTEM DESIGN

The generated flat spectrum ASE source is then spectrally sliced using an Arrayed Waveguide Grating (AWG). The bandwidth and frequency spacing of the AWG is optimized to achieve better slicing. Spectrally spliced 16- and 32-channel configurations from the AWG with a frequency spacing of 100 GHz (0.8 nm) are used. The frequency grid and interval is defined in ITU-T Recommendation G.694.1 (Forrester et al., 1991). The DWDM-PON system has a passive optical link which does not contain any active components. Figure 2 shows the DWDM-PON system, which consists of a Central Office (CO), an ODN and end users. The information from the central office is transmitted to the end users via the ODN. The central office consists of ASE source generation and OLTs, and each end user has one Optical Network Terminal (ONT). A standard ITU-T G.652 single-mode optical fiber is used as transmission link.

A Long-reach PON (LPON) system is an attractive solution for long-haul transmission. The concept of an LPON using a single-sided mid-span extender has been of research interest since the 1990s (Forrester et al., 1991). Researchers have focused on increasing the length of PON networks via mid-span optical amplifiers (Suzuki et al., 2007) and transponders (Davey et al., 2006). The ITU recommended and standardized the concept of using optical amplifiers in ITU-T Recommendation G.984.6. Both the DWDM-PON and long-reach PON systems consist of an ASE broadband source, an AWG as both multiplexer (MUX) and demultiplexer (DEMUX), an optical fiber link, and a PIN diode at the receiver side. Figure 2 represents the WDM-PON and long-reach PON systems. The performance investigation of this system is analyzed using a generated broadband ASE signal source. The signal source is then passed into a gain-flattening filter in order to obtain a flat spectrum. Depending upon the bandwidth of the filter used after the gain-flattening filter, the number of channels used is determined as 16 or 32. Then the

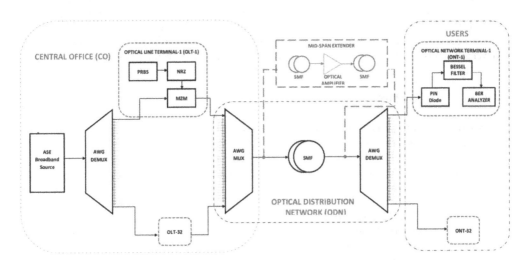

Figure 2. Schematic design of 32-channel PON and long-reach PON systems.

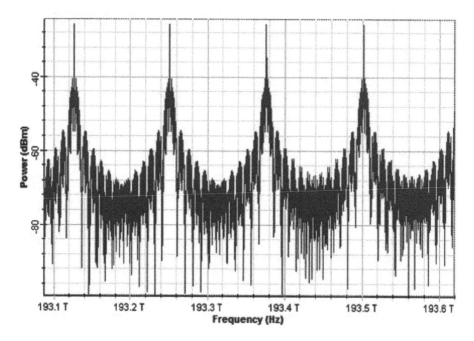

Figure 3.   Modulated signal using ASE source.

extracted spectrum is passed through the AWG to generate the spectrally sliced individual channels.

Using this AWG, the spectrally sliced optical channels with dense frequency spacing of 100 GHz (0.8 nm) are obtained. Then the spectrally sliced channels are transmitted to the OLTs at the central office. The responsibility of each OLT is to modulate the data with the spectrally spliced individual channels. Each OLT comprises a Pseudo-Random Binary Sequence (PRBS), Non-Return to Zero (NRZ) signal generator and Mach–Zehnder Modulator (MZM). The bit sequence generated from the PRBS is sent to the NRZ driver where electrical NRZ pulses are generated, which are then modulated by the MZM. Figure 3 represents the modulated signal using the ASE source. The modulated signals are then multiplexed and transmitted to Standard Single-Mode optical Fiber (SSMF). The ODN, comprised of optical fiber, acts as an intermediary for the transmission between the central office and the end users (i.e. from OLT to ONT). The ONT on the receiver side consists of a PIN photodiode and filter to eliminate unwanted noise.

4   RESULTS AND DISCUSSION

The performance of the DWDM-PON system described above was investigated for bit rates of both 2.5 and 10 Gbps over 32 channels by utilizing the flat ASE spectrum source.

The performance of each end-user link is evaluated on the basis of the Bit Error Rate (BER) obtained. Based on ITU G.984.2 recommendations, the BER value for fiber optical transmission systems with a 2.5 Gbps data rate is specified as less than 10^{-10}. Using the gain-flattening filter at the end of the EDFA, an ASE light source of -20 dBm with a flat spectrum starting from 192.0 THz with a frequency spacing of 100 GHz is chosen. In a 16-channel 2.5 Gbps system, a BER of 9.54e-014 and a Q-factor of 7.35 are obtained for a fiber span length of 15 km. The same system is investigated for 32 channels and a BER of 2.49e-09 and a Q-factor of 5.84 are obtained for a fiber span length of 3 km. Because of the increase in channels and bit rate, the transmission distance decreases.

Figure 4 shows the eye diagrams for a single user channel obtained at the receiver side for 2.5 Gbps/16 channels and 2.5 Gbps/32 channels for both PON and LPON systems. The same 16/32-channel system setup is also analyzed for a bit rate of 10 Gbps without any active element in the transmission link and worse BER and Q-factor values are obtained (see Figure 5). These results demonstrate that the acceptable transmission of higher bit rates in DWDM-PON systems is impossible. In order to successfully achieve long-distance transmission in DWDM-PON systems, optical amplifiers can be used to create long-reach DWDM-PON systems. The 2.5/10 Gbps 32-channel system was analyzed with a single optical amplifier and it was found that transmission over fiber lengths of 105 km/68 km can be obtained with BERs of 1.58e-010 and 1.18e-012, respectively.

In addition, different modulation formats, such as Return to Zero (RZ), Non-Return to Zero (NRZ), Modified Duo-binary Return to Zero (MDRZ), Carrier-Suppressed Return to Zero (CSRZ) and duo-binary modulation format, for a system of 10 Gbps 32-channel DWDM long-reach PON were investigated. For our system, it was shown that the NRZ modulation format performed best, and an improved BER and Q-factor were obtained. Furthermore, to achieve

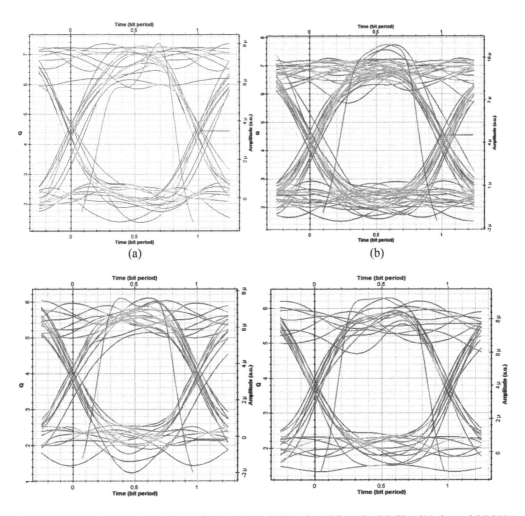

Figure 4.   Eye diagrams for: (a) 2.5 Gbps/16-channel PON for 32 km; (b) 2.5 Gbps/16-channel LPON for 110 km; (c) 2.5 Gbps/32-channel PON for 8 km; (d) 2.5 Gbps/32-channel LPON for 105 km.

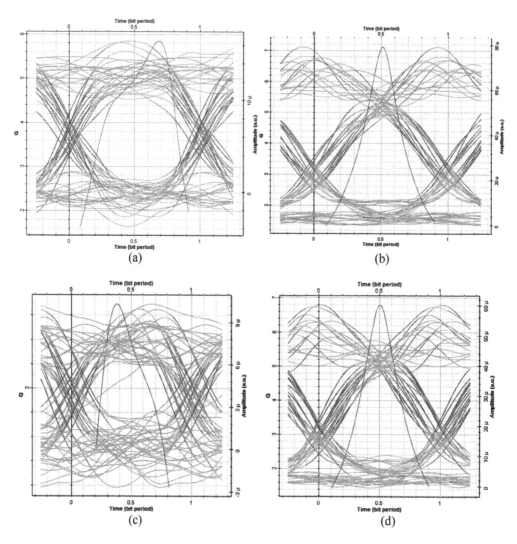

Figure 5. Eye diagrams for (a) 10 Gbps/16-channel PON for 3 km; (b) 10 Gbps/16-channel LPON for 75 km; (c) 10 Gbps/32-channel PON for 5 km; (d) 10 Gbps/32-channel LPON for 68 km.

a better transmission distance with better BER, a dispersion-compensation module can be inserted in the transmission link. This module consists of Dispersion-Compensation Fiber (DCF) as well as an amplifier. Dispersion compensation can be achieved by using DCF with a dispersion of 80 ps/km/nm. By effective use of this module, long-distance transmission can be achieved.

ACKNOWLEDGMENT

The authors (D Shanmuga sundar and Marcos Flores Carrasco) would like to thank the Conicyt FONDECYT (Fondo Nacional de Desarrollo Científico y Tecnológico) Project No. 3180089 and Millennium Nucleus MULTIMAT for funding and support.

REFERENCES

ACTS Project AC050: Photonics Local Access Network (PLANET).

Cauvin, A., Tofanelli, A., Lorentzen, J., Brannan, J., Templin, A., Park, T. & Saito, K. (2006). Common technical specification of the G-PON system among major worldwide access carriers. *IEEE Communications, 44*(10), 34–40.

Choi, B.H. & Lee, S.S. (2011). The effect of AWG-filtering on a bidirectional WDM-PON link with spectrum-sliced signals and wavelength-reused signals. *Optics Communications, 284*(24), 5692–5696.

Davey, R.P., Healey, P., Hope, I., Watkinson, P., Payne, D.B., Marmur, O., Ruhmann, J. & Zuiderveld, Y. (2006). DWDM reach extension of a GPON to 135 km. *Journal of Lightwave Technology, 24*(1), 29–32.

El-Sahn, Z.A., Mathlouthi, W., Fathallah, H., LaRochelle, S. & Rusch, L.A. (2010). Dense SS-WDM PON over legacy PONs: Smooth upgrade of existing FTTH networks. *Journal of Lightwave Technology, 28*(10), 1485–1495.

Forrester, D.S., Hill, A.M., Lobbett, R.A. & Carter, S.F. (1991). 39.81 Gbit/s 43.8 million way WDM broadcast network with 527 km range. *Electronics Letters, 27.*

Kaneko, S., Kani, J., Iwatsuki, K., Ohki, A., Sugo, M. & Kamei, S. (2006). Scalability of spectrum-sliced DWDM transmission and its expansion using forward error correction. *Journal of Lightwave Technology, 24*(3), 1295–1301.

Lee, K. Lim, D.S., Jhon, M.Y., Kim, H.C., Ghelfi, P., Nguyen, T., Poti, L. & Lee, S.B. (2012). Broadcasting in colorless WDM-PON using spectrum-sliced wavelength conversion. *Optical Fiber Technology, 18*(2), 112–116.

Lin, C. (Ed.). (2006). *Broadband optical access networks and fiber-to-the-home: Systems technologies and deployment strategies.* New York, NY: Wiley.

Spolitis, S. & Ivanovs, G. (2011). Extending the reach of DWDM-PON access network using chromatic dispersion compensation, IEEE Swedish Communication Technologies Workshop (Swe-CTW 2011), Art. No. 6082484, pp. 29–33.

Suzuki, K., Foukada, Y., Nesset, D. & Davey, R. (2007). Amplified gigabit PON systems. *Journal of Optical Networking, 6*(5), 422–433.

Advances in Optoelectronic Technology and Industry Development – Jose & Ferreira (eds)
© 2020 Taylor & Francis Group, London, ISBN 978-0-367-24634-1

A comparative selection of the low-loss optical fibers designed for FTTH networks

Faramarz E. Seraji, Ali Emami & Davood Ranjbar Rafi
Optical Communication Group, Iran Telecom Research Center, Tehran, Iran

ABSTRACT: Fiber-to-the-home (FTTH) technology provides the ability to transmit telecommunication data at a maximum level. In order to achieve this technology, fibers with the lowest bending and splice losses are required. In this paper, the losses of bending and splice joints in single mode optical fibers have been reviewed with a variety of designs, including holey fibers, nanoscale optical fibers, and microstructured optical fibers. By comparing the losses of these fibers, we chose the best fiber with the lowest losses in designing FTTH that is used for two purposes of reducing costs and improving optical network performances. These fibers have advantages such as low bending radius, wavelength and mode field diameter, which can be effective in minimizing bending losses.

Keywords: Holey optical fibers, microstructured optical fibers, nanoscale optical fibers, bending losses, bending radius, splice losses, FTTH

1 INTRODUCTION

For Fiber-To-The-Home (FTTH) networks, optical fibers should be insensitive to bending in order to minimize installation costs and improve the network performance (Wagner et al., 2006). For installation of optical fibers at home, the used fibers must have a minimum bending loss less than 0.1 dB/km (Chen et al., 2008).

In the early years of the use of optical fibers in optical communication systems, the bending loss has been investigated theoretically and experimentally by several researchers (Waluyo et al., 2018). It was shown that when light-wave propagates to a bent section of optical fiber, the field distribution changes considerably such that beside a transition loss, there is macro-bending loss, as well (Gambling et al., 1979). It was further revealed that bending loss of a single-mode fiber was oscillating with wavelength and bending radius (Wang et al., 2005; Zendehnam et al., 2010).

At a particular wavelength, the bending loss has a considerable value. For each type of Single-Mode Fiber (SMF), the bending loss of the LP11 and LP01 modes at a wavelength are significant when the normalized frequency (V-number) ranges between 2 and 4.4, and 1 and 2.4, respectively (Bayuwati & Waluy, 2018).

Optical power loss caused by bending is not desired in optical fiber communication systems with operating wavelengths between 1300 – 1600 nm. Based on the ITU-T Rec. G.652, the maximum bending loss at 1550 nm for 100 turns of fiber with bending radius of 37.5 mm is 0.50 dB (ITU Rec. ITU-T G.652, 2010).

A method based on the perturbation theory is reported to calculate the bending losses of individual modes of Few-Mode Fibers (FMFs), which is applied for trench-assisted fibers (Zheng et al., 2016). It is shown that changing the distance of the trench-core, for each order of mode, there is a minimum bending loss, which can be used for fiber optimization. It is found that the bending performance of parabolic-index FMFs is better than that of step-index FMFs with fixed core radius and cutoff wavelength.

In order to reduce the bending losses in optical fibers used in FTTH networks, the design of Holey Optical Fibers (HOFs) has been proposed that are referred to as Microstructure Optical Fibers (MOFs). In this design, with a bending radius of 5 mm, the mode field diameter (MFD) was 9.3 µm at a wavelength of 1550 nm and cutoff wavelength of 1100 nm. The measured bending loss and splice loss were obtained less than 0.011 dB/turn and 0.08 dB, respectively (Shinohara, 2005; Bing et al., 2005; Tsuchida & Saito, 2005).

One of the fibers for this technology is a fiber of nanoscale structure, insensitive to bending, with ultra-low bending loss is designed and manufactured. Bending radius of 5 mm at an operating wavelength of 1550 nm, yielded a bending loss of less than 0.1 dB/trun (Li et al., 2008). Other optical parameters of the designed fiber fully comply with the standard communications grade single-mode fibers.

As the FTTH networks deployment is increasing, optical fiber cables are required to be handled easily with less construction space. Under this circumstance, an SMF providing small bending radius is strongly needed (Tsuchida & Saito, 2005).

A novel single-mode HOF bending-insensitive with doped core and two layers of air-holes with different diameters is proposed. The fiber has an MFD of 9.3 µm at operating wavelength of 1550 nm and a cutoff wavelength below 1100 nm, showed a bending loss of 0.011 dB/turn at 1550 nm for a bending radius of 5 mm and a low splice loss of 0.08 dB per fusion when spliced to a conventional SMF (Tsuchida & Saito, 2005).

Another novel MOF for use in FTTH, with 12 air-holes in its cladding is reported (Luo et al., 2014). The bending losses of this novel MOF at 2 mm and 5 mm bending radius is reported less than 0.1 dB and 0.049 dB, respectively, at 1550 nm wavelength. The splice loss between the novel MOF with standard SMF is measured as 0.12 dB. This presented novel MOF technology is a strong candidate to achieve low-loss bending-insensitive solutions in comparison to conventional communication optical fibers.

In this paper, these reported optical fibers, designed for use in FTTH networks have been studied and compared. According to reported measurements, the minimum bending losses for each optical fiber, depending on different parameters such as bending radius, MFD, and operating wavelength, are determined and then the best designed fiber suitable for use in FTTH technology will be introduced.

2 INVESTIGATING THE DESIGN OF OPTICAL FIBERS INSENSITIVE TO BENDING

Increasing demands for high capacity and low cost of optical communication systems have increased the importance of high-volume optical fibers with flexible structures. This is achieved when optical fibers can be used with a minimum bending radius.

Meanwhile, conventional single-mode optical fibers have a shortcoming of having a minimum bending radius of 5 mm, which is required for an acceptable fiber functionality in FTTH networks. Therefore, there should be a lot of space required in using these types of fibers. Thus, for the FTTH networks, optical fibers that are not sensitive to bending are preferred for better network performances (Tsuchida & Saito, 2005; Li et al., 2008; Luo et al., 2014; Matsuo et al., 2004; Islam et al., 2015; Kozlov, 2012; Chauhan et al., 2016; Krähenbühl et al., 2010).

2.1 *Holey optical fiber structure*

For applications in the FTTH networks, optical fibers should have low bending losses, shorter cutoff wavelengths at higher modes, and MFD that are consistent with conventional optical fibers. In recent years, Photonic Crystal Fibers (PCFs), also known as holey optical fibers (HOF) and microstructured optical fiber (MOF), have attracted the attention of researchers for their unique features that are not found in conventional optical fibers.

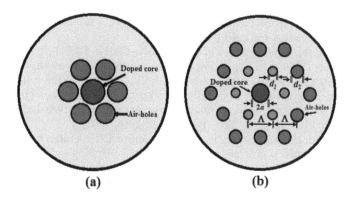

Figure 1. (a) A holey optical fiber with one layer of air-holes, (b) with two layers of air-holes (Tsuchida & Saito, 2005).

The HOF has at least one layer of air-holes in the cladding region that surrounds the central area named as core. In Figure 1, two types of HOFs are shown, with one layer (Figure 1a) and two layers (Figure 1b) of six- and twelve-air-holes ring arranged in a triangular lattice form in the cladding, where a denotes the core radius, d_1 and d_2 are diameters air-holes of the inner and the outer layers, respectively, and Λ represents the pitch of the holes in the cladding region (Tsuchida & Saito, 2005). It is noted that d_1 and d_2 are not necessarily equal, but are uniform in fiber cross section. The concentration of germanium in the core region is 37% by weight.

In this design, the cutoff wavelength is a definite number and is equal to 1650 nm in order to determine the allowable bending radius for the minimum bending losses, since for the HOFs, the bending losses are higher at longer wavelengths. If the total bending loss for 10 turns is less than 0.5 dB at 1650 nm, then the radius of each turn will be defined as the allowable bending radius.

Here, the feasibility of splicing a conventional SMF to the HOF with minimum losses will be investigated. The HOF is chosen for splicing because when the air-holes of the HOF are collapsed at the splice point, its MFD coincides with that of SMF, thus will cause fewer splice losses during the installation process.

The bending losses are the function of the MFD, which on decrease will reduce the bending losses. In Figure 1b, with increasing the value of d_1, the cutoff wavelength would increase, as well. By keeping d_1 and Λ constant and only changing the value of d_2, one can control the cutoff wavelength and bending losses without altering the values of the MFD (Tsuchida & Saito, 2005).

When the HOF fiber is spliced to the SMF fiber, the splice point would experience some losses at the joint. To determine the minimum required MFD of the HOF, in order to have total splice losses L_s less than 0.3 dB at 1550 nm operating wavelength, we use the following well-known expression as (Tsuchida & Saito, 2005; Marcuse, 1977):

$$L_s = -20 \, \log_{10} \frac{2 \, w_{SMF} \, w_{HOF}}{w_{SMF}^2 + w_{HOF}^2} \tag{1}$$

where w_{SMF} and w_{HOF} are MFDs of the SMF and the HOF, respectively.

By inserting the constant value of w_{SMF} and the varying values of HOF from 5 μm to 8 μm in Equation 1, we obtain the splice losses in a given range of MFD of HOFn (Tsuchida & Saito, 2005). In Figure 2a, the HOF splice losses are calculated in terms of the MFD and plotted for $w_{SMF} = 2.5$ μm.

We note that for values of $MFD_{HOF} > MFD_{SMF}$, the splice loss between them increases exponentially, as depicted in Figure 2a, while for $MFD_{HOF} < MFD_{SMF}$, the trend of splice loss variations will attain a decreasing nature, as shown in Figure 2b.

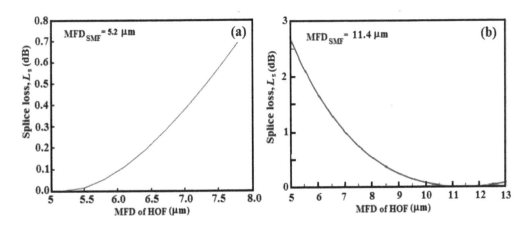

Figure 2. The splice loss as a function of MFD of the HOF for (a) $MFD_{SMF} = 5.2$ μm and (b) $MFD_{SMF} = 11.4$ μm (Tsuchida & Saito, 2005).

The result of calculating the splice loss in terms of the MFD for SMF (or HOF) indicates that the MFD has a direct relation with the losses of the splice point between two jointed optical fibers. Therefore, for splicing SMF with HOF, by considering a suitable range for MFD, it is possible to minimize the splice loss between these two types of fibers during installation work.

As it is stated, in terms of losses, HOFs can meet the needs of FTTH networks. The HOF fibers are suitable options for splicing to the SMF. The designs of germanium-doped core HOFs with typically two layers of air-holes of different diameters in the cladding region are reported. It is shown via numerical and experimental results that it is possible to design an HOF with following parameters: an MFD of 9.3 μm at wavelength of 1550 nm, a small bending loss of 0.011 dB/turn at bending radius of 5 mm, and a cutoff wavelength below 1100 nm by choosing a suitable core radius, hole pitch, and air-hole diameters, as shown in Figure 3 (Tsuchida & Saito, 2005). The characteristic parameters of the fabricated HOF at 1550 nm are presented in Table 1.

Worth to note that the HOF bending losses are much less than that of the SMF fiber (Tsuchida & Saito, 2005; Marcuse, 1977). The average measured splice loss at 1550 nm is 0.08 dB per fusion-splice between two spliced fibers (Hasegawa et al., 2003).

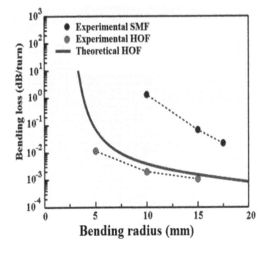

Figure 3. The characteristic of bending loss as function of bending radius of HOF and SMF fibers at 1550 nm wavelength (Tsuchida & Saito, 2005).

Table 1. Characteristic parameters of the fabricated HOF at 1550 μm.

Loss (dB/km)	MFD (μm)	Cutoff λ(nm)	Dispersion (ps/nm.km)	Bending loss (dB/turn) Bend dia. (5 mm)	Bend dia. (20 mm)
2.3	9.3	1100	21.8	0.011	0.002

2.2 Nanoscale optical fiber structure

Nanoscale Optical Fiber (NOF) is a new structure of the HOF, which consists of a germanium-doped core and a nanoscale ring layer in the cladding. This ring consists of condensed nanometer air-holes arranged along the fiber with a diameter of a few hundred nanometers, as shown in Figure 4. This ultra-low bending loss SMF with nanometer-sized ring is designed and manufactured (Li et al., 2008; Wu et al., 2006; Matsuo et al., 2007; de Montmorillon et al., 2006).

Bending losses of the NOF were measured using five turns on different bending radii. Figure 5 compares typical bending losses in terms of bending radius for the NOFs, standard SMF and trench fiber designs at operating wavelength of 1550 nm. The results

Figure 4. Nanoscale optical fiber insensitive to bending (Li et al., 2008).

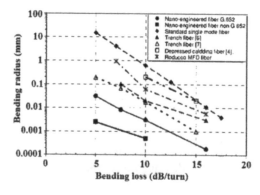

Figure 5. Comparison of bending performance of NOF with standard SMF and trench fibers (Li et al., 2008).

104

Table 2. Optical characteristics of the NOF (Luo et al., 2014).

λ (nm)	MFD (μm)	Dispersion (ps/nm.km)	Loss (dB/km)	Cable λ_{cutoff} (nm)	Zero dispersion λ (nm)	Dispersion slope (ps/nm².km)
1310	8.7	-	0.34	1225	1315	0.09
1550	9.7	17.6	0.20	1225	1315	0.09

reveal that the bending losses of the NOF is about 500 times better than the standard SMF, and 6–10 times better than the trench fibers. The average bending loss at 5 mm radius is 0.03 dB/turn.

Other optical parameters of the designed NOF are fully compatible with the standard communications grade. The obtained results are consistent with the OVD fabrication process. The optical characteristics of this type of fiber at wavelengths of 1310 and 1550 nm are shown in Table 2. Minimization of bending and splice losses are one of the advantages of the NOF; this is the feature which is important for FTTH networks (Luo et al., 2014).

2.3 *Microstructured optical fiber with 12 air-holes*

Microstructured optical fiber (MOF) with new features and outstanding optical components can produce insensitive bending properties. The MOF bending losses at a bending radius of 2.5 mm can be less than 0.1 dB, according to the published reports. In comparison with conventional fibers, MOFs have a relatively high splice loss.

Meanwhile, MOF has manufacturing problems that appear while controlling the shapes of the air-holes. These problems make it hard to achieve the lowest cost of production, as the production process of MOF is more complex than conventional optical fiber. Therefore, it is necessary to design a new structure that is easy to construct and less affected by changes of parameters during production in order to obtain the desired applications for this type of fiber in the FTTH networks (Luo et al., 2014).

In Figure 6, the structural design of the MOF with a ring of 12 air-holes located in the cladding is illustrated, which is prepared by the drill method combined with the conventional preform fabrication processes.

With various turning wraps at different radii, the bending losses are calculated at certain wavelengths and the results are presented in Table 3. The bending loss of the MOF at 2.5 mm

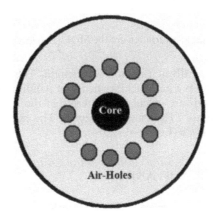

Figure 6. The MOF with a ring of 12 air-holes in the cladding of the fiber (Luo et al., 2014).

Table 3. The parameters of the MOF with a ring of 12 air-holes (Luo et al., 2014).

Bending radius (mm)	Wrap (Turn)	λ (nm)	Bending loss (dB)
2	1	1550	0.89
		1625	0.86
3	10	1550	0.73
		1625	0.044
5	1	1550	0.049
		1625	0.047

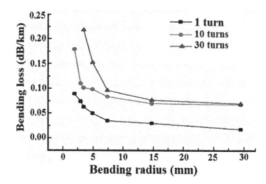

Figure 7. Bending losses in terms of bending radius for different turns at 1550 nm (Matsuo et al., 2004).

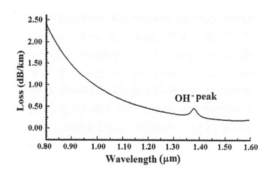

Figure 8. The calculated spectral loss variations of the MOF (Luo et al., 2014).

bending radius is less than 0.1 dB at 1550 nm wavelength. The splice loss of the MOF with G.652D fibers is 0.12 dB. This is a good achievement for avoiding bending losses.

The bending losses of MOF are calculated in terms of the bending radius with 1, 10, 30 turns, and are plotted in Figure 7 (Matsuo et al., 2004). The calculated spectral loss variatins of the fiber illustrated in Figure 8 (Luo et al., 2014).

3 COMPARISON OF HOF, NOF, AND MOF

We have compared the characteristic parameters of three selected fibers with different structures that are used in FTTH networks. A summary of the comparison is presented in Table 4 (Tsuchida & Saito, 2005; Li et al., 2008; Luo et al., 2014). The parameters values indicate that

Table 4. The comparison of characteristic parameters of three selected fibers with different structures.

No.	Fiber types	Bending loss (dB/turn)	Splice loss (with SMF dB)
1	HOF	0.011	0.080 (Tsuchida & Saito, 2005)
2	NOF	≤ 0.1	0.033 (Li et al., 2008)
3	MOF/12 air-holes	0.049	0.120 (Luo et al., 2014)
	Bending radius = 5 mm, Wavelength = 1550 nm		

in terms of bending loss, fiber No. 2 is better selection, while with respect to splice loss with the standard SMF, one may prefer fiber No. 1.

We further note that more experimental investigations could be done by considering the influence of characteristic structural parameters such as air-hole diameter d, the hole pitch Λ, and the ratio d/Λ using large-mode area and small-mode-area photonic crystal fibers (Seraji & Kasiri, 2015).

4 CONCLUSION

The present work has reviewed single-mode optical fiber with different structures that may be used in FTTH optical networks in conjunction with insensitivity to bending losses. In this paper, we have compared three selected fibers by considering the effects of their structures on bending losses, prominently, with respect to bending radius and operating wavelength.

By considering the results of the analyses, all three types of optical fibers at bending radii of 5 mm and operating wavelength of 1550 nm have different bending losses. The HOF, as compared to two other types of fibers, has lower bending loss of 0.011 dB/turn at the given operating wavelength of 1550 nm. Therefore, among these optical fibers, it can be the most suitable option for use in FTTH networks in terms of bending loss.

In continuation of the present investigation in conjunction with the reduction of the bending loss of single-mode fibers used in FTTH networks, we suggest that the influences of characteristic parameters, such as air-hole diameter d, the hole pitch Λ, and the ratio d/Λ to be further studied for photonic crystal fibers of large-mode and small-mode areas.

ACKNOWLEDGMENT

The authors are thankful to the Research Council of Iran Telecom Research Center for allotment of the project with Code No. 440950900 to the optical communication group in Communication Technology Dept.

REFERENCES

Bayuwati, D. & Waluy, T.B. 2018. Macro-bending Loss of Single-mode Fiber beyond Its Operating Wavelength", TELKOMNIKA, Vol.16, No.1, pp.142~150.
Bing, Y. et al. 2005. Low-loss Holey Fiber", Hitachi Cable Rev., No.24.
Chauhan, J.D. et al. 2016. For Fiber-to-the-home (FTTH) Architecture using Passive Optical Network (PON): A Review," Int'l. J. Mod. Trends Eng. and Research, Vol. 3, Iss. 4, pp. 673–677.
Chen, D.Z. et al. 2008. Requirements for bend insensitive fibers for Verizon's FiOS and FTTH applications", OFCNFOEC2008, San Diego, CA, Feb. 24–28.
de Montmorillon, L.-A. et al. 2006. Bend-optimized G.652D compatible trench-assisted single-mode fibers," Proc. 55th IWCS/Focus, Nov. 2006, pp. 342–347.
Gambling, W.A. et al. 1979. Curvature and microbending losses in single-mode optical fibres", Opt. Quant. Electron, Vol.11, pp. 43–59.

Hasegawa, T. et al. 2003. Bending-insensitive single-mode holey fiber with SMF-compatibility for optical wiring applications," in proceedings of European Conf. Opt. Commun. (ECOC2003), pap. We2.7.3, Rimini, Italy.

Islam, R. et al. 2015. Bend-Insensitive and Low-Loss Porous Core Spiral Terahertz Fiber," IEEE Photon. Technol. Lett., Vol. 27, Iss. 21, pp. 2242–2245.

ITU Rec. ITU-T G.652, 2010. Characteristics of a Single-Mode Optical Fibre and Cable. Geneva.

Kozlov, V. 2012. Corning bend insensitive optical fibers for elevated temperature applications", Proc. IEEE Avionics, Fiber-Optics and Photon. Technol. Conf. (AVFOP), 11-13 Sept., 2012. DOI:10.1109/AVFOP.2012.6344038.

Krähenbühl, R. et al. 2010. Compatibility of Low Bend Single mode Fibers", R+D Technology Fiber Optics Division, HUBER+SUHNER AG, 9100, Herisau, Switzerland.

Li, M.J. et al. 2008. Ultra-low Bending Loss Single-Mode Fiber for FTTH", IEEE J. Lightwave Technol., Vol. 27, Iss. 3, pp. 376–382.

Luo, W.Y. et al. 2014. Low-loss Bending-insensitive Micro-structured optical fiber for FTTH", Proc. 61st IWCS Conf., International Wire & Cable Symp., Nov. 11–14, pp. 454–457.

Marcuse, D. 1977. Loss analysis of single-mode fiber splices", Bell Syst. Tech, J. Vol. 56, No. 5, pp. 703–718.

Matsuo, S. et al. 2004. Bend-insensitive and low-splice-loss optical fiber for indoor wiring in FTTH", OFC'2004, Opt. Fiber Commun. Conf., Los Angeles, Calif, US. 22 Feb. 2004, pap. Th13.

Matsuo, S. et al. 2007. Design optimization of trench index profile for the same dispersion characteristics with SMF," OFCNFOEC'2007, pap. JWA2.

Seraji, F.E. & Kasiri, S. 2015. Optimization of Macrobending Loss in Small and Large Mode Area, Photonic Crystal Fibers", Open Acc. Lib. J., 2: e2269. http://dx.doi.org/10.4236/oalib.1102269.

Shinohara, H. 2005. Broadband Access in Japan: Rapidly Growing FTTH Market", IEEE Communications Magazine, Vol. 43, Iss. 9, pp. 72–78, Sept 2005.

Tsuchida, Y. & Saito, K. 2005. Design and characterization of single-mode holey fibers with low bending losses", Opt. Express, Vol. 13, Issue 12, pp. 4770–4779.

Wagner, R.E. et al. 2006. Fiber-based broadband-access deployment in the United States", J. Lightwave Technol., vol. 24, no. 12, pp. 4526–4540.

Waluyo, T.B. et al. 2018. The effect of macro-bending on power confinement factor in single mode fibers", Journal of Physics: Conf. Series 985, IOP Publishing.

Wang, Q. et al. 2005. Theoretical and experimental investigations of macro-bend Losses for standard single mode fibers", Opt. Express, *Vol.* 13, pp. 4476–4484.

Wu, F. et al. 2006. A new G.652D, zero water peak fiber optimized for low bend sensitivity in access networks," in IWCS 2006, Providence, RI, Nov. 12–15.

Zendehnam, A. et al. 2010. Investigation of bending loss in a single-mode optical fibre", Pramana–J. Phys, Vol. 74, Issue 4, pp 591–603.

Zheng, X. et al. 2016. Bending losses of trench-assisted few-mode optical fibers", Appl. Opt., Vol. 55, Issue 10, pp. 2639–2648.

Advances in Optoelectronic Technology and Industry Development - Jose & Ferreira (eds)
© 2020 Taylor & Francis Group, London, ISBN 978-0-367-24634-1

Enhancement of fidelity of quantum teleportation in a non-Markovian environment

Yanliang Zhangy
College of Software, Jishou University, Zhangjiajie, P. R. China

Xingqi Wu
School of Mechanical Electronic and Information Engineering, China University of Mining and Technology, Beijing, P. R. China

ABSTRACT: We investigate quantum teleportation of a single unknown qubit state for the dissipative situation in which both qubits of the entangled channels that are formed by the pure and mixed maximally entangled states are subjected to local structured reservoirs. We consider the dissipative factors in Markovian and non-Markovian regimes on the average fidelity of the teleportation by solving the quantum master equation. It is shown that the fidelity of teleportation decreases exponentially over time in Markovian environments and attenuates oscillatorily in non-Markovian ones. In addition, when the non-Markovian and detuning conditions are satisfied simultaneously, the fidelity increases as the detuning increases. Even with a lower purity of entanglement, fidelity can be preserved effectively at comparatively high levels.

1 INTRODUCTION

Entanglement is one of the most striking features of quantum mechanics; it is also regarded as an important physical resource in quantum communication, computation, dense coding, and so on. As one of the possible applications of quantum information theory, teleportation is universally acknowledged as the most attractive quantum state transmission protocol, which allows an unknown quantum state to be transmitted between two parties (usually dubbed Alice and Bob) by using the resources of quantum entanglement and classical communication. In the seminal work of teleportation, Bennett and Brassard (1984) proposed a scheme for transporting an unknown single-body quantum state via single copy of the maximally entangled state as quantum channel. Later on, teleportation has been extensively investigated both experimentally and theoretically, ranging not only from two-level states to high-dimensional state regimes, but also from discrete variable to continuous variable domains (Bennett et al., 1993; Barrett et al., 2004; Pan et al., 2001; Wagner & Clemens, 2009; Jin et al., 2005). However, the practical implementation of any quantum information protocol has to face the problem of the unavoidable coupling of the quantum system with its environment. Indeed, real systems can never be perfectly isolated from the surrounding world (Zhang et al., 2010). It is therefore important to understand the impact of the coupling with a noisy environment on the stability of quantum protocols.

Bennett et al. (1993) noted that the quantum channel which is non-maximally entangled may reduce the fidelity of teleportation and/or the range of states that can be accurately teleported. Oh et al. (2002) have investigated quantum teleportation through noisy quantum channels by solving analytically and numerically a master equation in Lindblad form. Hao et al. (2006) have considered the effects of amplitude damping in quantum noise channels on average fidelity of quantum teleportation using Kraus representation. Furthermore, Kumar and Pandey (2003) evaluated the fidelities of teleportation under a variety of decoherence conditions and different entangled channels; they found that a generic coupling to environment using the singlet state for the entangled channel yields the highest fidelity in noisy conditions. Recently, Hu et al. (2010) investigated the standard teleportation in a dissipative environment and showed the quality of teleportation was determined by both the entanglement and the purity of the channel state, and the highest fidelity can be effectively ensured by optimal matching of these.

In this paper, we investigate quantum teleportation of a single-qubit state for the situation in which both qubits of the entangled channel are subjected to local structured reservoirs. We study the teleportation protocol in the case of the entanglement resources being subject to a dissipative environment with weak coupling (Markovian) and strong coupling (non-Markovian), which usually emerge in cavity Quantum Electrodynamics (cavity-QED) systems (Madsen et al., 2011; Kaer et al., 2010). In contrast to the bulk of previous studies, we concentrate on the dynamics of an open system coupled to a structured reservoir, in which the teleportation fidelity exhibits oscillatory behavior. It is found that if we merely modify the center frequency of the laser cavity to make the detuning large enough, then the fidelity of teleportation can be preserved and promoted.

2 THE FIDELITY OF QUANTUM TELEPORTATION WITH A NON-IDEAL QUANTUM CHANNEL

Alice and Bob share a pair of maximally entangled states, such as one of the four Bell states described in the following form:

$$|\psi_{Bell}^{0,3}\rangle = \frac{1}{\sqrt{2}}(|00\rangle \pm |11\rangle), |\psi_{Bell}^{1,2}\rangle = \frac{1}{\sqrt{2}}(|01\rangle \pm |10\rangle) \tag{1}$$

It is easy to implement the transmission of the unknown quantum state $|\psi_{in}\rangle = \cos\theta|0\rangle + e^{i\phi}\sin\theta|1\rangle$ through joint Bell measurement and classical communication. In the unknown quantum state $|\psi_{in}\rangle$ the $\theta(\in [0, \pi])$ and $\phi(\in [0, 2\pi])$ are unknown to Bob and even to Alice. In the dissipative environments, the teleportation is described by density matrices of the total initial state $|\psi\rangle_{in}\langle\psi| \otimes \rho^c(t)$ and the teleported output state is given by

$$\rho_{out}^m = \sum_{k=0}^{3}\langle\psi_{Bell}^{k\oplus m}|\rho^c(t)|\psi_{Bell}^{k\oplus m}\rangle \otimes \sigma^k|\psi_{in}\rangle\langle\psi_{in}|\sigma^{\dagger k} \tag{2}$$

where m ($m = 0, 1, 2, 3$) represents the m-th Bell state that is chosen as quantum channel and $k \oplus m$ is the summation modulus 4. The time-dependent density operator $\rho^c(t) \in H_a \otimes H_b$ is the noisy quantum channel evolved from the m-th entangled Bell state and $|\psi_{in}\rangle\langle\psi_{in}| \in H_{in}$ is an unknown input state to be teleported. $|\psi_{Bell}^i\rangle$ ($i = 0, 1, 2, 3$) are the familiar four maximally entangled Bell states in Equation 1 denoting joint Bell measurement by Alice, and σ^k ($k = 0, 1, 2, 3$) are the standard Pauli operators denoting local operations by Bob. In this process, Bob's local operations just correspond to the results of Alice's joint Bell measurements that are announced through classical communication. Because the $|\psi_{in}\rangle$ is uncertain, it is more beneficial to calculate the average fidelity (the fidelity averaged over all possible measurement outcomes m for Alice and all possible pure input states $|\psi_{in}\rangle$ on the Bloch sphere) to quantify the teleportation process. This average fidelity is:

$$F^m = \chi_0^m + \frac{1}{3}\left[\chi_1^m + \chi_2^m + \chi_3^m\right] \tag{3}$$

where $\chi_k^m = \chi_{k\oplus m}$ with $\chi_i = \langle\psi_{Bell}^i|\rho^c(t)|\psi_{Bell}^i\rangle$, which are given by $\chi_{0,3} = (1/2)(\rho_{11} + \rho_{44} \pm \rho_{14} \pm \rho_{41})$, $\chi_{1,2} = (1/2)(\rho_{22} + \rho_{33} \pm \rho_{23} \pm \rho_{32})$.

We consider two two-level entangled systems (quantum entangled channel for teleportation shared by Alice and Bob) interacting independently with their corresponding vacuum reservoirs at zero temperature and with no other interaction in the whole system. The single *qubit* + *reservoir* dynamics can be described by the following Hamiltonian (Breuer & Petruccione, 2002):

$$H = \hbar\omega_0\hat{a}^\dagger\hat{a} + \hbar\sum_{k=1}^{N}\omega_k\hat{b}_k^\dagger\hat{b}_k + \hbar\sum_{k=1}^{N}g_k(\hat{a}\hat{b}_k^\dagger + \hat{a}^\dagger\hat{b}_k) \tag{4}$$

where ω_0 is the transition frequency, $\hat{a}^\dagger$ and $\hat{a}$ are the system creation and annihilation operators, the index k labels the field modes of the reservoir with frequencies ω_k, $\hat{b}_k^\dagger$ and $\hat{b}_k$ are the field k-th mode creation and annihilation operators, and g_k the coupling constants. Under the Hamiltonian (Equation 4), the single-qubit dynamics can be represented by:

$$|0\rangle|\mathbf{0}\rangle \Rightarrow |0\rangle|\mathbf{0}\rangle,$$

$$|1\rangle|\mathbf{0}\rangle \Rightarrow |\psi_t\rangle = C_0(t)|1\rangle|\mathbf{0}\rangle_r + C(t)|0\rangle|\mathbf{1}\rangle \tag{5}$$

where $|\mathbf{0}\rangle = \Pi_{k=1}^{N}|0_k\rangle$ denotes the N-mode vacuum reservoir and $|\mathbf{1}\rangle = \frac{1}{C(t)}\sum_{k=1}^{N}C_k(t)|1_k\rangle$ with $|1_k\rangle$ being the state of the N-mode reservoir with only one excitation in the k-th mode. $C_0(t)$ and $C(t)$ are the times-dependent probability amplitudes related to the structure of the reservoir.

The Hamiltonian (Equation 4) can conveniently represent a model for damping of a qubit formed by the excited and ground electronic state of a controllable two-level atom interacting with a reservoir formed by the quantized modes of a high-Q cavity. At zero temperature, this Hamiltonian represents one of the few open quantum systems amenable to an exact solution. Here, we take the spectral distribution of an electromagnetic field inside an imperfect cavity supporting a mode detuned by Δ from the qubit transition frequency ω_0. This Lorentzian spectral density can have the form (Dalton et al., 2001):

$$J(\omega) = \frac{1}{2\pi}\frac{\gamma_0\lambda^2}{(\omega_0 - \omega - \Delta)^2 + \lambda^2} \tag{6}$$

where λ is the width of the distribution that describes the decay rate into the reservoir and the reservoir correlation time τ_R is the inverse of λ, the weight γ_0 relates to the decay of the excited state of the atom in the Markovian limit of flat spectrum, and the relaxation time scale τ_S is the inverse of γ_0. The time evolution of state $|\psi_t\rangle$ in Equation 5 is governed by the Schrödinger equation $i\hbar\frac{\partial}{\partial t}|\psi_t\rangle = H|\psi_t\rangle$ with the Hamiltonian (Equation 4); the exact probability amplitude is obtained as:

$$C_0(t) = e^{-\frac{1}{2}(\lambda - i\Delta)t}\left[\cosh\left(\frac{\Omega t}{2}\right) + \frac{\lambda - i\Delta}{\Omega}\sinh\left(\frac{\Omega t}{2}\right)\right],$$

$$C(t) = \sqrt{1 - C_0^2(t)}, \tag{7}$$

where $\Omega = \sqrt{(\lambda - i\Delta)^2 - 2\gamma_0\lambda}$. It can be noted that with $\Delta = 0$ the character of qubit and reservoir interactions can be divided into weak and strong coupling regimes (Bellomo et al., 2007). When $\gamma_0 < \lambda/2$, it means the weak coupling regime, that is, $\tau_S \gg \tau_R$, the behavior of the system is Markovian and irreversible decay occurs, controlled by γ_0. When $\gamma_0 > \lambda/2$, it means the strong coupling regime, that is, $\tau_R \gg \tau_S$. The factor $\lambda/2\gamma_0$ can characterize the degree of Markovian and non-Markovian effects.

Without loss of generality, it is assumed that the initial entangled channel we consider involves Werner mixed states in the following forms:

$$\rho^{\psi_{Bell}^{0,3}} = r|\psi_{Bell}^{0,3}\rangle\langle\psi_{Bell}^{0,3}| + \frac{1-r}{4}I,$$

$$\rho^{\psi_{Bell}^{1,3}} = r|\psi_{Bell}^{1,2}\rangle\langle\psi_{Bell}^{1,2}| + \frac{1-r}{4}I \tag{8}$$

where r is the purity if the initial entangled states range from 0 to 1. For $r = 0$, the Werner states become totally mixed and they reduce to the Bell pure states, $|\psi_{Bell}^i\rangle$ in the case of $r = 1$. Because the $|\psi_{Bell}^0\rangle$ and $|\psi_{Bell}^3\rangle$, $|\psi_{Bell}^1\rangle$ and $|\psi_{Bell}^2\rangle$ have the same structures excepting symbols of some entries, we only give the two reduced density matrices of them in the dissipative environments as follows (Zhang et al., 2012):

$$\rho_{11}^{\psi_{Bell}^0}(t) = \frac{r+1}{4}|C_0(t)|^4, \; \rho_{14}^{\psi_{Bell}^0}(t) = \rho_{41}(t)\frac{r}{2}|C_0(t)|^2,$$

$$\rho_{22}^{\psi_{Bell}^0}(t) = \rho_{33}(t) = \frac{1-r}{4}|C_0(t)|^2 + \frac{r+1}{4}|C_0(t)|^2|C(t)|^2,$$

$$\rho_{44}^{\psi_{Bell}^0}(t) = 1 - [\rho_{11}(t) + \rho_{22}(t) + \rho_{44}(t)], \; \rho_{ij}^{\psi_{Bell}^0}(t) = 0, \; i, j = others \tag{9}$$

and

$$\rho_{11}^{\psi_{Bell}^1}(t) = \frac{1-r}{4}|C_0(t)|^4, \; \rho_{23}^{\psi_{Bell}^1}(t) = \rho_{32}(t) = \frac{r}{2}|C_0(t)|^2,$$

$$\rho_{22}^{\psi_{Bell}^1}(t) = \rho_{33}(t) = \frac{r+1}{4}|C_0(t)|^2 + \frac{1-r}{4}|C_0(t)|^2|C(t)|^2,$$

$$\rho_{44}^{\psi_{Bell}^1}(t) = 1 - [\rho_{11}(t) + \rho_{22}(t) + \rho_{44}(t)], \; \rho_{ij}^{\psi_{Bell}^1}(t) = 0, \; i, j = others \tag{10}$$

For the dissipative channels of Equations 9 and 10 with $r = 1$, we plot the dynamic of the fidelity as a function of the spectral width of the coupling λ and evolution time t in the Figures 1 and 2, respectively.

In the Markovian regime, we set $\gamma_0 = 0.02$ and the value of λ is ranged from 0.04 to 0.2, and the degree of Markovian effect ranges from 1 to 5. In a non-Markovian regime, we set $\gamma_0 = 1.6$ and the value of λ is ranged from 0.01 to 0.4, and the degree of non-Markovian effect ranges from 3.125×10^{-3} to 0.125. From Figures 1 and 2, the fidelity decays exponentially and converges to a fixed value asymptotically in a finite time in the Markovian environment, and the degree of Markovian effect has little impact on the fidelity. In the non-Markovian regime, the fidelity diminishes instantaneously under the disturbance of non-Markovian noise, but with the course of time, it periodically decreases with a damping of its revival amplitude, and the fidelity is much more sensitive to the degree of the non-Markovian effect. The smaller the value of $\lambda/2\gamma_0$ is, the more intense the oscillation that the fidelity exhibits, and the memory effects of the non-Markovian reservoir lead to an appearance of revival of average fidelity.

Compared with Figures 1 and 2, the quantum channel $|\psi_{Bell}^0\rangle$ is more robust than $|\psi_{Bell}^1\rangle$, in which the fidelity remains at 2/3 for $|\psi_{Bell}^0\rangle$ and 1/3 for $|\psi_{Bell}^1\rangle$ for a comparatively long time. However, the memory effect of fidelity with $|\psi_{Bell}^1\rangle$ is stronger than that with $|\psi_{Bell}^0\rangle$ in

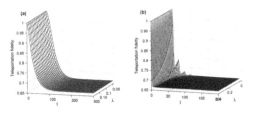

Figure 1. The average fidelity as a function of λ and t when the initial quantum channel is $|\psi^0_{Bell}\rangle$ and the system is subject to: (a) A Markovian environment; (b) a non-Markovian environment.

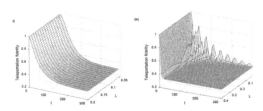

Figure 2. The average fidelity as a function of λ and t when the initial quantum channel is $|\psi^1_{Bell}\rangle$ and the system is subject to: (a) A Markovian environment; (b) a non-Markovian environment.

a non-Markovian regime. The reason is mainly that the local structured reservoirs on the two qubits of the quantum channel for $|\psi^0_{Bell}\rangle$ and $|\psi^1_{Bell}\rangle$ are not symmetric.

For the dissipative channels of Equations (9) and (10) with $0 < r < 1$, we have plotted the average fidelities as a function of evolution time t and the purity of entanglement of quantum channel r in Figures 3 and 4. The fidelities decrease with the decreasing of purity of entanglement of quantum channel at a fixed time, especially at the beginning of interaction with its environment.

Compared with Figures 3 and 4, at a fixed purity of entanglement of initial quantum channel, the memory effects of fidelity with the channel $|\psi^1_{Bell}\rangle$ are more obvious than that with channel $|\psi^1_{Bell}\rangle$, but the final fidelity value is just about $1/3$, which is much smaller than that with channel $|\psi^0_{Bell}\rangle$. Moreover, it is easy to find that no matter how much the purity of the initial entanglement is, the fidelities converge towards their own fixed values with the evolution times, respectively. The correlation information of quantum state is, in general, derived from quantum entanglement and classical correlation. When a pure entangled state is considered, its entanglement accounts for the whole of the correlation information. On the other hand, for a mixed state, the correlation information may also be due to classical correlation.

Another intriguing aspect is how the fidelity preservation is influenced by the values of $\lambda/2\gamma_0$. In Figure 5, when the detuning conditions are satisfied simultaneously, the fidelity can effectively be enhanced. For the memory effects of a non-Markovian

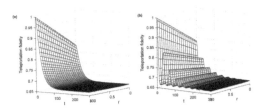

Figure 3. The average fidelity as function of r and t when initial quantum channel is $\rho^{\psi^0_{Bell}}$ and the system subject to Markovian environment (a) and non-Markovian environment (b).

113

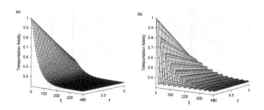

Figure 4. The average fidelity as function of r and t when initial quantum channel is $\rho^{\psi^1_{Bell}}$ and the system subject to Markovian environment (a) and non-Markovian environment (b).

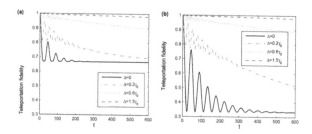

Figure 5. [Color online] The average fidelity in a non-Markovian environment when the quantum channels, with $r = 1$, are: (a) $\rho^{\psi^0_{Bell}}$; (b)$\rho^{\psi^1_{Bell}}$.

environment, when the detuning is increasing, the fidelity of teleportation is preserved at comparatively high levels even approaching its maximum value of 1. When detuning $\Delta = 1.5\gamma_0$, the fidelity of teleportation almost never decreases. From this respect, in this dissipative environment, the detuning produces a postponing effect on fidelity decrease. This means that, in effective quantum information processing time, the quantum communication protocol sustains a comparatively higher fidelity, which is of crucial importance in quantum information processing. With this result, in a non-Markovian dissipative environment one can merely modify the center frequency of the laser cavity to make the detuning large enough that the fidelity of teleportation can be preserved and promoted. Comparing Figures 5a and 5b, although the average fidelity of teleportation by using channel $|\psi^0_{Bell}\rangle$ can be kept at a higher level than by using channel $|\psi^1_{Bell}\rangle$ in the same dissipative environment, the enhancement of average fidelity by modifying the quantity of detuning is much more easily achieved than when the channel is formed by state $|\psi^1_{Bell}\rangle$. Furthermore, in the situation for the smaller purity of mixed entangled channel, the average fidelity can be promoted by detuning in a non-Markovian environment. Figure 6 illustrates that, even when the purity of the mixed entangled channel is 0.2, the average fidelity can be promoted above the level of the pure entangled channel without detuning.

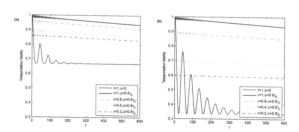

Figure 6. [Color online] The average fidelity in a non-Markovian environment when the quantum channels, with $r = 0.2$, are: (a) $\rho^{\psi^0_{Bell}}$; (b) $\rho^{\psi^1_{Bell}}$.

4 CONCLUSIONS

In summary, by solving the master equation governing the dynamics of qubits in an open system, we have derived analytically the expressions of the pure and mixed two-qubit maximally entangled system with a single reduced density matrix in the presence of Markovian and non-Markovian dissipative environments (weak and strong coupling regimes of an open quantum system). We presented a detailed investigation of the teleportation abilities of a single unknown qubit by using the resource of two-qubit entanglement as a quantum channel in a dissipative environment. By examining the dynamics of average fidelities of teleportation with respect to evolution time t and spectral width of the coupling λ, we show that the average fidelity is decreasing exponentially over time in a Markovian regime, and attenuating oscillatorily in a non-Markovian regime. Furthermore, the capability for teleportation is not sensitive to λ in a Markovian regime, while it is very different to that in a non-Markovian regime. λ characterizes the degree of non-Markovian effects when γ_0 is fixed, in which the memory effects of a non-Markovian reservoir lead to an apparent revival of average fidelity. Meanwhile, the quantum channel formed by $|\psi_{Bell}^0\rangle$ is much more robust than that formed by $|\psi_{Bell}^1\rangle$, in which the former remains stable at the value of 2/3 while the latter is at 1/3 for a comparatively long time. For the mixed entangled channel, the decreasing of purity mainly causes the reduction of average fidelity both in Markovian and non-Markovian dissipative environments but does not change the influence of Markovian and non-Markovian effects on the capability for teleportation.

In a non-Markovian environment, we also presented the phenomenon of robust fidelity preservation when the interactions between the entangled channel and dissipative environments, modeled as structured reservoirs with spectral density $J(\omega)$, are non-resonant. And we have shown that non-resonant conditions can be used to prevent degradation of teleportation fidelity. That is to say, we can promote the value of teleportation fidelity in a non-Markovian environment. Even in the lower purity of mixed entangled channels, the robust fidelity of teleportation can be enhanced and preserved efficiently.

Furthermore, the spectral distribution in Equation (6) has been widely employed in quantum optics, and was recently used in several studies. Because strong coupling between matter and light has been available experimentally in some systems, on a short timescale it could, in principle, be observed and even manipulated. As a result, non-Markovian effects would become more and more important in the exploration of quantum information processing under real experimental conditions. Our study of teleportation in the Markovian and non-Markovian dissipative environment will be useful in quantum information processing in realistic open systems.

ACKNOWLEDGMENT

The work was supported by the National Natural Science Foundation of China under Grant No.11464015. General Science Research Foundations of Education Department of Hunan Province under Grants No.18C0579 and No.11A096.

REFERENCES

Barrett, M.D., Chiaverini, J., Schaetz, T., Britton, J., Itano, W.M., Jost, J.D., Knill, E., Langer, C., Leibfried, D., Ozeri, R. Wineland, D.J. (2004). Deterministic quantum teleportation of atomic qubits. *Nature, 429*, 737.

Bellomo, B., Lo Franco, R. & Compagno, G. (2007). Non-Markovian effects on the dynamics of entanglement. *Phys Rev Lett, 99*, 160502.

Bennett, C.H. & Brassard, G. (1984). Quantum cryptography: public-key sistribution and tossing. *Proceedings of the IEEE International Conference on Computers, Systems and Signal Processing, Bangalore*, India. (p. 175)

Bennett, C.H., Brassard, G., Crepeau, C., Jozsa, R., Peres, A. & Wootters, W.K. (1993). Teleporting an unknown quantum state via dual classical and Einstein-Podolsky-Rosen channels. *Phys Rev Lett, 70,* 1895.

Breuer, H.P. & Petruccione, F. (2002). The theory of open quantum systems. Oxford, UK: Oxford University Press.

Dalton, B.J., Barnett, S.M. & Garraway, B.M. (2001). Theory of pseudomodes in quantum optical processes. *Phys Rev A, 64,* 053813.

Hao, X., Zhang, R. & Zhu, S.Q. (2006). Average fidelity of teleportation in quantum noise channel. *Commun Theor Phys (Beijing, China), 45,* 802.

Hu, X.Y., Gu, Y., Gong, Q.H. & Guo, G.C. (2010). Noise effect on fidelity of two-qubit teleportation. *Phys Rev A, 81,* 054302.

Hu, M.L. (2011). Teleportation of the one-qubit state in decoherence environments. *J Phys B: At Mol Opt Phys, 44,* 025502.

Jin, L.H., Jin, X.R. & Zhang, S. (2005). Teleportation of a two-atom entangled state with a thermal cavity. *Phys Rev A, 72,* 024305.

Kaer, P., Nielsen, T.R., Lodahl, P., Jauho, A.-P. & Mørk, J. (2010). Non-Markovian model of photon-assisted dephasing by electron-phonon interactions in a coupled quantum-dot-cavity system. *Phys Rev Lett, 104,* 157401.

Kumar, D. & Pandey, P.N. (2003). Effect of noise on quantum teleportation. *Phys Rev A, 68,* 012317.

Li, Y.L. & Fang, M.F. (2011). High entanglement generation and high fidelity quantum state transfer in a non-Markovian environment. *Chin Phys B, 20,* 100312.

Madsen, K.H., Ates, S., Lund-Hansen, T., Löffler, A., Reitzenstein, S., Forchel, A. & Lodahl, P. (2011). Observation of non-Markovian dynamics of a single quantum dot in a micropillar cavity. *Phys Rev Lett, 106,* 233601.

Nielsen, M.A. & Chuang, I.L. (2000). Quantum computation and quantum information. Cambridge, UK: Cambridge University Press.

Oh, S., Lee, S. & Lee, H.W. (2002). Fidelity of quantum teleportation through noisy channels. *Phys Rev A, 66,* 022316.

Pan, J.W., Daniell, M., Gasparoni, S., Weihs, G. & Zeilinger, A. (2001). Experimental demonstration of four-photon entanglement and high-fidelity teleportation. *Phys Rev Lett, 86,* 4435.

Wagner, R., Jr. & Clemens, J.P. (2009). Performance of a quantum teleportation protocol based on temporally resolved photodetection of collective spontaneous emission. *Phys Rev A, 79,* 042322.

Xiao, X., Fang, M.F., Li, Y.L., Zeng, K. & Wu, C. (2009). Robust entanglement preserving by detuning in non-Markovian regime. *J Phys B: At Mol Opt Phys, 42,* 235502.

Zhang, Y.L., Zhou, Q.P., Kang, G.D., Zhou, F. & Wang, X.B. (2012). Remote state preparation in non-markovian environment. *Int J Quantum Inf, 10,* 1250030.

Zhang, Y.J., Zou, X.B., Xia, Y.J. & Guo, G.C. (2010). Different entanglement dynamical behaviors due to initial system-environment correlations. *Phys Rev A, 82,* 022108.

Advances in Optoelectronic Technology and Industry Development – Jose & Ferreira (eds)
© 2020 Taylor & Francis Group, London, ISBN 978-0-367-24634-1

A 2×2 optical switch based on semiconductor optical amplifier cross-gain modulation technology

Shaohua Zhou, Zhengwei Qi, Deqiang Ding & Dapeng Deng
School of Information and Communication, National University of Defense Technology, Xi'an, P. R. China

ABSTRACT: A 2×2 optical switch structure was designed and investigated based on semiconductor optical amplifier cross-gain modulation technology. The 2×2 optical switch structure not only can realize optical-controlled switch function, but also can amplify the exchanged optical signal. The correctness of the 2×2 optical switch was verified by simulation on an OptiSystem simulation platform. Simulation shows that the signal to be connected was amplified from -10dBm to 16dBm, and the other signal was suppressed, and the extinction ratio was up to 26dB, by the control of 15dBm control optical pulse, and the switch speed exceeds 20Gbit/s.

Keywords: semiconductor optical amplifier, cross-gain modulation, optical switch, OptiSystem, switch speed

1 INTRODUCTION

The all-optical switch is a key component to solve the electronic bottleneck problem and realize optical information processing in optical communication, optical computers and All-Optical Networks (AONs). At present, the proposed all-optical switch structure is basically realized by taking advantage of the high-speed nonlinear effect of some optical media, such as the Semiconductor Optical Amplifier (SOA), Electrical Absorption Modulator (EAM), optical waveguide device and, photonic crystal. Although the optical switch structure based on the photonic crystal, optical waveguide device and electrical absorption modulator has the advantages of small power consumption and easy integration, the insertion loss is very high, especially for the electrical absorption modulator. As the authors Evankow and Thompson (1988), Williams et al. (2005), Aw et al. (2007a) and cheng et al. (2013) described, the switch structure based on the semiconductor optical amplifier not only has the advantages of small size, easy integration and fast response time, but also can amplify the switched optical signal, and it has attracted much attention in recent years.

For example, Huh et al. (2007) realized the optical switch function with an all-optical switch triode based on the cross-gain modulation effect of SOA; Hui et al. (2009) designed and researched a multi-wavelength optical packet switching structure based on SOA, which realized a dynamic range of about 16dB; and Tanaka et al. (2009) designed and integrated an 8:1 SOA gated optical switch structure with on-off gain exceeding 14dB and extinction ratio exceeding 70dB. Wang et al. (2012) designed a 4 × 4 SOA optical switch with a dynamic input power range of 12dB and insertion loss of 1dB, and realized a 16 × 16 SOA optical switch array by three-stage cascade of the 4 × 4 SOA optical switch; but, due to the accumulation of amplified spontaneous emission noise and saturated absorption distortion, the insertion loss of the optical switch increased to 2.5dB. Cheng et al. (2015) monolithically integrated and researched a 64 × 64 optical switching structure with insertion loss as small as 1dB. Ding (2017) designed a 16 × 16 optical switch structure using cascaded 4 × 4 dilated hybrid MZI-SOA optical switches, with insertion loss less than 1dB and a dynamic input power range of 15dB. Li et al. (2018) designed and researched a 1×2 wavelength conversion optical switch system based on the Four-Wave Mixing (FWM) effect of SOA by multi-level SOA wavelength conversion and amplification; the gain was greater than 25dB, the bit error rate was improved from 10^{-5} to 10^{-18}, and the Q factor was improved from 4.16 to 8.17 compared with the first-level switch structure.

Based on the above research, in order to further reduce the insertion loss, and improve the gain and extinction ratio of the optical switch, this paper designed a 2×2 gain optical switch based on the Cross-Gain Modulation (XGM) technology of SOA. The amplification factor of four semiconductor optical amplifiers was controlled by the input control light pulse, which realized the switch function by amplifying or suppressing the input optical signal with the absence or presence of the control light pulse. The signal was amplified from -10dBm to 15dBm, the other signal was suppressed, the extinction ratio was greater than 25.5dB, and the switching speed exceeded 20Gbit/s, when the power of the control light pulse was 12dBm, by simulation on an OptiSystem simulation platform.

2 2×2 OPTICAL SWITCH BASED ON SOA-XGM TECHNOLOGY

2.1 *SOA cross-gain modulation technology*

As Tadashi and Takaaki. (1990), Annetts et al. (1997), Moerk et al. (2003), Kim and Chuang (2006), and Meuer et al. (2008) described, in a semiconductor optical amplifier, gain saturation effect is the phenomenon where the small signal amplification factor is large and the large signal amplification factor is small. The relationship between input optical power P_{in} and output optical power Pout of SOA is as follows:

$$P_{out} = G * P_{in} \qquad (1)$$

where G is the gain (amplification factor), and the relationship between G and the small signal gain G_0 is as follows:

$$G = G_0 * e^{(1-G)*\frac{P_{tot}}{P_{sat}}} \qquad (2)$$

where P_{tot} is the sum of the total power of input signal P_{in} and the power of spontaneous emission noise P_{ASE}. The formula shows that the greater the input optical signal power P_{in} is, the lower the gain (amplification factor) G is. When the input two optical signals wavelengths are λ_1 and λ_2, then:

$$P_{tot} = P_{in}(\lambda_1) + P_{in}(\lambda_2) + P_{ASE} \qquad (3)$$

where $P_{in}(\lambda_1)$ and $P_{in}(\lambda_2)$ are the optical power of λ_1 and λ_2, respectively. If the spontaneous emission noise power P_{ASE} is ignored, then $P_{tot} = P_{in}(\lambda_1) + P_{in}(\lambda_2)$. When the power difference of the two optical signals is very large, for example $P_{in}(\lambda_1) \gg P_{in}(\lambda_2)$, then the gain (amplification factor) G is mainly dependent on $P_{in}(\lambda_1)$:

$$G(\lambda_1) = G_0 * e^{(1-G)*\frac{P_{in}(\lambda_1)}{P_{sat}}} \qquad (4)$$

If the power of input optical signal $P_{in}(\lambda_1) > \ln G_0 * P_{sat}/(G-1)$, the gain (amplification factor) $G(\lambda_1) < 1$, the output power of the other optical signal λ_2 is as follows:

$$P_{out}(\lambda_2) = G(\lambda_1) * P_{in}(\lambda_2) < P_{in}(\lambda_2) \qquad (5)$$

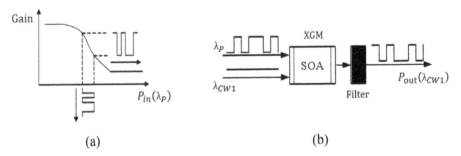

(a)   (b)

Figure 1.   Saturation gain effect in SOA and light-controlled switch structure and waveform based on XGM.

This shows that the gain (amplification factor) $G(\lambda_2)$ is controlled by another power of input optical signal $P_{in}(\lambda_1)$. Figure 1a shows the saturated gain effect in SOA and Figure 1b shows the switch structure and waveform of the light-controlled switch based on cross-gain modulation. When the control optical power $P_{in}(\lambda_P)$ is large, the gain G will be small because of the saturated gain effect of SOA, and so the output optical power $P_{out}(\lambda_{cw1})$ will be small; when the control optical power $P_{in}(\lambda_P)$ is small, the gain G will be large, and so the output optical power $P_{out}(\lambda_{cw1})$ will be large.

2.2   2×2 optical switch structures based on SOA-XGM

The 2×2 optical switch structure based on SOA-XGM is shown in Figure 2, which consists of four semiconductor optical amplifiers, four filters, four 3 dB beam splitters and six beam combiners, and can be divided into four SOA switches S1, S2, S3 and S4. The port1 and port2 are the input port, and the port3 and port4 are the output port. The signal of wavelength λ_P is the control laser, which controls the gain of the four semiconductor optical amplifiers SOA1, SOA2, SOA3 and SOA4. The function of the filter1 and filter3 is to allow the laser signal of wavelength λ_{cw1} to pass through and filter out the laser signals of other wavelengths; in the same way, the function of the filter2 and filter4 is to allow the laser signal of wavelength λ_{cw2} to pass through and filter out the laser signals of other wavelengths. If the gain of the semiconductor optical amplifiers SOA1, SOA2, SOA3 and SOA4 are G_1, G_2, G_3 and G_4 respectively, which are controlled by the laser signal of wavelength λ_P, and if the insertion loss such as of the filter, beam splitters and beam combiners is ignored, then the output optical power of port3 and port4 are as follows:

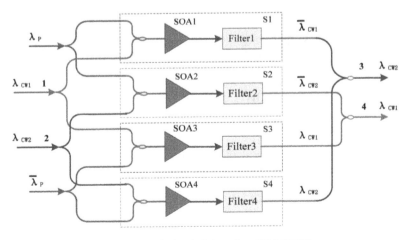

Figure 2.   Schematic diagram of 2 × 2 optical switch based on SOA-XGM.

119

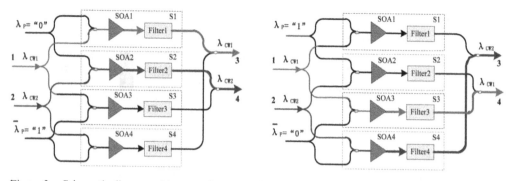

Figure 3. Schematic diagram of input and output relationship of 2×2 optical switch: (a) 'bar' status; (b) 'cross' status.

$$P_{out3} = \frac{1}{2}(G_1 * P_{\lambda_{cw1}} + G_4 * P_{\lambda_{cw2}}) P_{out4} = \frac{1}{2}(G_2 * P_{\lambda_{cw2}} + G_3 * P_{\lambda_{cw1}}) \qquad (6)$$

(1) 'Bar' status. As shown in Figure 3a, the control optical power of wavelength λ_P is small enough in SOA1 and SOA2, but large enough in SOA3 and SOA4, so the gain (amplification factor) G_1 and G_2 will be large, and G_3 and G_4 will be small because of the saturation gain effects. Thus the output optical powers of port3 and port4 are as follows:

$$P_{out3} = \frac{1}{2}(G_1 * P_{\lambda_{cw1}} + G_4 * P_{\lambda_{cw2}}) \approx \frac{1}{2}G_1 * P_{\lambda_{cw1}}$$

$$P_{out4} = \frac{1}{2}(G_2 * P_{\lambda_{cw2}} + G_3 * P_{\lambda_{cw1}}) \approx \frac{1}{2}G_2 * P_{\lambda_{cw2}} \qquad (7)$$

This shows that the laser signal of wavelength λ_{cw1}(port1) is magnified and output from port3, and the laser signal of wavelength λ_{cw2}(port2) is magnified and output from port4.

(2) 'Cross' status. As shown in Figure 3b, the control optical power of wavelength λ_P is large enough in SOA1 and SOA2, but small enough in SOA3 and SOA4, so the gain (amplification factor) G_1 and G_2 will be small because of the saturation gain effects, G_3 and G_4, will be large. Thus the output optical powers of port3 and port4 are as follows:

$$P_{out3} = \frac{1}{2}(G_1 * P_{\lambda_{cw1}} + G_4 * P_{\lambda_{cw2}}) \approx \frac{1}{2}G_4 * P_{\lambda_{cw2}}$$

$$P_{out4} = \frac{1}{2}(G_2 * P_{\lambda_{cw2}} + G_3 * P_{\lambda_{cw1}}) \approx \frac{1}{2}G_3 * P_{\lambda_{cw1}} \qquad (8)$$

This shows that the laser signal of wavelength λ_{cw1}(port1) is magnified and output from port4, and the laser signal of wavelength λ_{cw2}(port2) is magnified and output from port3.

3 PERFORMANCE SIMULATION

As shown in Figure 4, the 2 × 2 optical switch based on SOA-XGM is set up on the OptiSystem simulation platform.

The wavelength of the control optical λ_P is set to 1,555 nm, and the power is set to 15dBm. The Mach-Zehnder interferometer is controlled by the NRZ (Non-Return-to-Zero) pulse code, which controls the power of the input control laser λ_P. After passing through a 3dB power beam splitter, two beams of a 12dBm control laser are injected into the corresponding SOA, which controls the gain of the semiconductor optical amplifier. The wavelength of the signal laser λ_{cw1} is set to 1,550 nm, and the power is set to -10dBm. After passing through a 3dB power beam splitter, two beams of -13dBm signal laser are injected into the

120

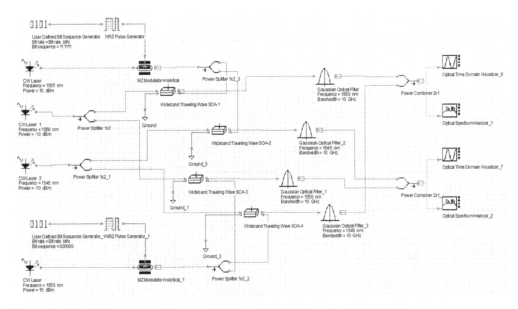

Figure 4. Simulation structure diagram of 2×2 optical switch based on SOA-XGM.

semiconductor optical amplifier SOA1 and SOA3, respectively; the wavelength of the signal laser λ_{cw2} is set to 1,545 nm, and the power is set to -10dBm. After passing through a 3 dB power beam splitter, two beams of -13dBm signal laser are injected into the semiconductor optical amplifier SOA2 and SOA4, respectively.

The simulation results of the 'bar' status are as shown in Figure 5 by the simulation on the OptiSystem simulation platform, where the current of the four semiconductor optical amplifiers is set to 600 mA. Figure 5a shows the optical signal spectrum of the output port3, which shows that the output optical power of the laser λ_{cw1} is 16dBm, the laser λ_{cw2} is -10.5dBm, and the extinction ratio of the optical switch is up to 26.5dB. Figure 5b shows the optical signal spectrum of the output port4, which shows that the output optical power of the laser λ_{cw1} is -12dBm, the laser λ_{cw2} is 16dBm, and the extinction ratio of the optical switch was up to 28 dB (extinction ratio EX = 16dBm-(-12dBm) = 28dB).

The simulation results of the 'cross' status are as shown in Figure 6, where the current of the four semiconductor optical amplifiers is set to 600 mA. Figure 6a shows the optical signal spectrum of the output port3, which shows that the output optical power of the laser λ_{cw1} is -11dBm, the laser λ_{cw2} is 16dBm, and the extinction ratio of the optical switch is up to 27dB. Figure 6b shows the optical signal spectrum of the output port4, which shows that the output optical power of the laser λ_{cw1} is 16dBm, the laser λ_{cw2} is -11dBm, and the

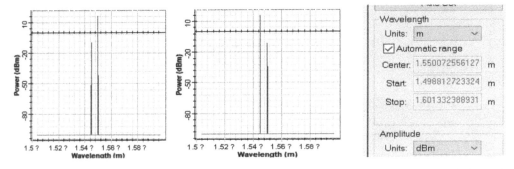

Figure 5. Output optical signal spectrum of port3 and port4 in 'bar' status: (a) Output optical signal spectrum of port3; (b) Output optical signal spectrum of port4.

121

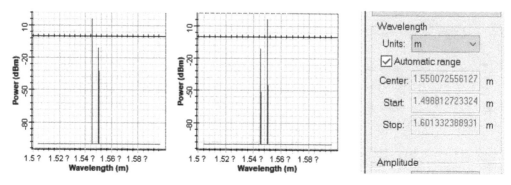

Figure 6.   Output optical signal spectrum of port3 and port4 in 'cross' status: (a) Output optical signal spectrum of port3; (b) Output optical signal spectrum of port4.

extinction ratio of the optical switch was up to 27dB (extinction ratio EX = 16dBm-(-11dBm) = 27dB).

3.1   *The relationship between the extinction ratio and the control optical power*

In 'cross' status, the relationship between the extinction ratio and the control optical power is as shown in Figure 7, which shows that the extinction ratio of the optical switch is always greater than or equal to 24dB. This changes little when the control light power is greater than 12dBm, but when the control light power is less than 12dBm, the extinction ratio decreases rapidly. For example, when the control light power is 9dBm, the extinction ratio decreases to 18.5 dB, which is because the effect of the cross-gain modulation is weakened due to the decrease of the control optical power. In 'bar' status, the relationship between the extinction ratio and the control optical power is basically the same as for the 'cross' status.

3.2   *The relationship between the optical switch performance and the wavelength of the control laser*

The relationship between the optical switch performance and the wavelength of the control laser is shown as in Figure 8, which shows that the output waveform diagrams are almost the same when the wavelength of the control laser is 1,540 nm, 1,547.5 nm and 1,555 nm, and so the optical switch performance is independent of the wavelength of the control laser.

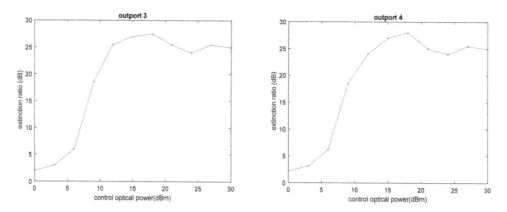

Figure 7.   The relationship diagram of the extinction ratio and the control optical power.

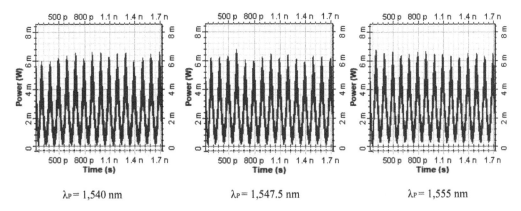

Figure 8. Relationship diagrams for optical switch performance and wavelength of the control laser.

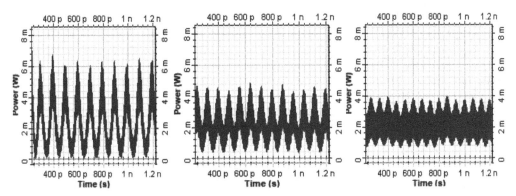

Figure 9. Simulation waveform diagrams at different optical switch speeds: (a) 20Gbit/s; (b) 25Gbit/s; (c) 30Gbit/s.

3.3 *Optical switch speed analysis*

The simulation waveform diagrams for different optical switch speeds are shown in Figure 9, which shows that the switch can work well at a speed of 20Gbit/s. However, the switch almost cannot work at a speed of 30Gbit/s; the signal quality starts to deteriorate at this speed because the carrier in the semiconductor optical amplifier cannot recover quickly enough with the increase of switch speed, and so the gain (amplification factor) G decreases.

4 CONCLUSIONS

The 2×2 optical switch structures based on SOA-XGM in this paper can work well at a speed of 20Gbit/s. Even at 25Gbit/s, the extinction ratio is more than 24dB when the control optical power is greater than 12dBm, and so it can widely be used in optical communications and optical networks in the future.

ACKNOWLEDGMENTS

This work is supported by the fund of the Science and Technology Program of Xi'an (201805049YD27CG33(1)).

REFERENCES

Annetts, J., Asghari, M. & White, I.H. (1997). The effect of carrier transport on the dynamic performance of gain-saturation wavelength conversion in MQW semiconductor optical amplifiers. *IEEE Journal of Selected Topics in Quantum Electronics, 3*(2), 320–329.

Aw, E.T., Lin, T., Wonfor, A., Penty, R.V., White, I.H. & Glick, M. (2007a). Multi-stage SOA switch fabrics: 4×40Gb/s packet switching and fault tolerance. In *Optical Fiber Communication Conference* (p. OThF2). Optical Society of America.

Aw, E.T., Wonfor, A., Glick, M., Penty, R.V. & White, I.H. (2007b). An optimized non-blocking SOA switch architecture for high performance Tb/s network interconnects. *Photonics in Switching, 15*–16.

Cheng, Q., Ding, M., Wonfor, A., Wei, J., Penty, R.V. & White, I.H. (2015). The feasibility of building a 64×64 port count SOA-based optical switch. In *2015 International Conference on Photonics in Switching* (pp. 22–25). IEEE.

Cheng, Q., Wonfor, A., Wang, K., Olle, V.F., Penty, R.V. & White I.H. (2013). Low-energy, scalable hybrid crosspoint switch design. In *Conference on Laser and Electro-Optics* (pp. CTu1L-3). Optical Society of America.

Ding, M., Wonfor, A., Penty, R.V. & White, I.H. (2017). Emulation of a 16×16 optical switch using cascaded 4×4 dilated hybrid MZI-SOA optical switches. In *2017 Optical Fiber Communications Conference and Exhibition*. IEEE.

Evankow, J. & Thompson, R.A. (1988). Photonic switching modules designed with laser-diode amplifiers. *IEEE Journal on Selected Areas in Communications, 6*(7), 1087–1095.

Ken Morito, Fujitsu Ltd., Morinosato-Wakamiya, Atsugi (2009). Optical switching devices based on semiconductor optical amplifiers. *2009 International Conference on Photonics in Switching* (pp. 15–19).

Huh, J.H., Homma, H., Nakayama, H. & Maeda, Y. (2007). All optical switching triode based on cross-gain modulation in semiconductor optical amplifier. *2007 Photonics in Switching* (pp. 19–22).

Hui, L., Imaizumi, H., Tanemura, T., Nakano, Y. & Morikawa, H. (2009). Experimental study on dynamic range of SOA switch for multi-wavelength optical packet switching. *2009 International Conference on Photonics in Switching* (pp. 15–19).

Kim, J. & Chuang, S.L. (2006). Small-signal cross-gain modulation of quantum-dot semiconductor optical amplifiers. *IEEE Photonics Technology Letters, 18*(23), 2538–2540.

Kim, J., Laemmlin, M., Meuer, C., Bimberg, D. & Eisenstein, G. (2008). Static gain saturation model of quantum-dot semiconductor optical amplifiers. *IEEE Journal of Quantum Electronics, 44*(7), 658–666.

Li, M.-X., Feng, H., Wang, H. & Xie, X.-P. (2018). Technical research on optical gain switch in optical packet switching network. *Acta Photonica Sinica, 47*(6).

Meuer, C., Kim, J., Laemmlin, M., Liebich, S., Capua, A., Eisenstein, G., ... Bimberg, D. (2008). Static gain saturation in quantum dot semiconductor optical amplifiers. *Optics Express, 16*(11), 8269–8279.

Antonella Bogoni, Luca Poti, Claudio Porzi, Mirco Scaffardi, Paolo Ghelfi, and Filippo Ponzini(2004) Modeling and Measurement of Noisy SOA Dynamics for Ultrafast Applications, *IEEE Journal of Selected Topics in Quantum* Electronics, 10(1), 197–205.

Takaaki, M. & Tadashi, S. (1990). Detuning Characteristics and Conversion Efficiency of Nearly Degenerate Four-Wave Mixing in a 1 S-pm Traveling-Wave Semiconductor Laser Amplifier, *IEEE Journal of Quantum Electronics, 26*, 20867.

Tanaka, S., Jeong, S.H., Yamazaki, S., Uetake, A., Tomabechi, S., Ekawa, M. & Morito, K. (2009). Monolithically integrated 8:1 SOA gate switch with large extinction ratio and wide input power dynamic range. *IEEE Journal of Quantum Electronics, 45*(9), 1155–1162.

Wang, K., Wonfor, A., Penty, R. & White I.H. (2012). Active-passive 4×4 SOA-based switch with integrated power monitoring. In *IEEE Optical Fiber Communication Conference and Exposition* (pp. 1–3). Optical Society of America.

Wang, K., Wonfor, A., Penty, R.V. & White I. (2013). Demonstration of cascaded operation of active-passive integrated 4 × 4 SOA switches with on-chip monitoring for power control and energy consumption optimization. In *Optical Fiber Communication Conference and Exposition and the National Fiber Optic Engineers Conference* (pp. 1–3). Optical Society of America.

Williams, K.A., Roberts, G.F., Lin, T., Penty, R.V., White, I.H., Glick, M. & McAuley, D. (2005). Integrated optical 2 × 2 switch for wavelength multiplexed interconnects. *IEEE Journal of Selected Topics in Quantum Electronics, 11*(1), 78–85.

Advances in Optoelectronic Technology and Industry Development – Jose & Ferreira (eds)
© *2020 Taylor & Francis Group, London, ISBN 978-0-367-24634-1*

A PSK quantum-noise randomized cipher simulation system model based on standard commercial devices

Yukai Chen, Tao Pu, Hua Zhou, Haiqin Shi, Haojun Tang & Yunkun Li
College of Communications Engineering, Army Engineering University of the People's Liberation Army (PLA), Nanjing, Jiangsu, China

Haisong Jiao*
College of Communications Engineering, Army Engineering University of the People's Liberation Army (PLA), Nanjing, Jiangsu, China
Troop 95841 of PLA, Jiuquan, Gansu, China

Hui Zhang
Troop 93117 of PLA, Nanjing, Jiangsu, China

ABSTRACT: On the basis of industry-standard devices, a Phase-Shift Keying (PSK) quantum-noise randomized cipher anti-interception transmission simulation system model with 10 Gb/s single-channel transmission over 198 km is simulated in this paper. The effect of the core parameters on the system transmission capability are estimated in consideration of bit error rate. The results show that the greater the transmission distance or rate, or the lower the average photon number, the worse the transmission performance becomes. However, the level number of M-ary signal (M) has little effect on the transmission performance, leading us to analyze the impact of M on system security. The eye diagram for an eavesdropper becomes worse with the increase of M. Therefore, legitimate users can achieve excellent security performance by advancing M without affecting transmission performance too much.

1 INTRODUCTION

The Quantum-Noise Randomized Cipher (QNRC) is a new anti-interception communication method combining the principles of quantum mechanics and classical stream ciphers. Due to Heisenberg's uncertainty principle, when the signal state reaches the mesoscale, unavoidable quantum noise causes photons' positional uncertainty, which covers up several constellation points, so that an eavesdropper ('Eve') is unable to obtain accurate ciphertext signals. However, owing to the shared key, legitimate users ('Alice' and 'Bob') can effectively avoid the influence of quantum noise, and achieve correct demodulation of information. At present, several implementation schemes for QNRC, such as Phase-Shift Keying (PSK), Intensity-Shift Keying (ISK), Quadrature Amplitude Modulation (QAM), and polarization modulation, are put forward. This study focuses on PSK, which receives less attention in all kinds of literature but has important significance. In 2005, online amplification QNRC fiber communication with single-channel data rates up to 650 Mbps over 200 km was realized by Yuen Group, which adopted two implementation schemes, of polarization and phase encoding (Corndorf et al., 2005). In 2009, a 256-PSK-QNRC system with 2.5 Gb/s single-channel transmission over 522 km was reported by NuCrypt LLC (Kanter et al., 2009). At the OFC2013 conference, Wang et al. (2013) demonstrated a Multi-PSK-QNRC scheme based on a self-coherent signal detection method. A continuous variable quantum secure communication system with real-time communication function was realized by the Zeng Guihua group of Shanghai Jiao Tong University (Dai, 2012). Meanwhile, through the investigation of QNRC experiments in recent years, the Y-00 protocol requires complex and high-speed digital-to-analog converter electronic chips, which suffer from high cost. Therefore, based on standard commercial devices, Jiao et al. (2018) conducted an 128-ISK-QNRC system experiment with single-channel data rates up to 2.5 Gb/s over 100 km.

A PSK-QNRC anti-interception transmission simulation system model is proposed in this paper. To the best of our knowledge, it is the first research into anti-interception transmission in PSK-QNRC on the basis of standard commercial devices. First, the principle of encryption of QNRC is demonstrated. Further, based on PSK-QNRC simulation system model, we analyze the influence of key parameters such as transmission rate, transmission distance, average photon number and M on transmission performance. Moreover, in light of the eye diagrams for Eve, the impact of M on system security performance is analyzed.

2   PRINCIPLE OF Y-00 ENCRYPTION

In this paper, an emerging method of physical cryptography is referred to as the Y-00 protocol. A secure key shared by Alice and Bob, which can be from the quantum key distribution channel, was extended as the key u to adapt to high-speed transmission systems. Then, data x was encrypted according to the encryption mapping function, namely:

$$m = f(x, u) = u + [x \oplus Pol(u)] \cdot 2^{|u|} \tag{1}$$

where $|u|$ is the length of the key, $M_b = 2^{|u|}$ is the number of ground states. Then, the cipher-text symbol m is modulated onto the mesoscopic state, obtaining the corresponding quantum state:

$$|\varphi(m)\rangle = \left| \alpha e^{im\frac{\pi}{M_b}} \right\rangle, \ m = 0, 1, \cdots 2M_b - 1 \tag{2}$$

where α is the coherent-state amplitude.

Figure 1 illustrates the encryption constellation of the Y-00 protocol. Quantum states of the same diameter have the same ground state. As for Bob owning the known key, the quantum state of the multi-binary system can be demodulated to the binary system. That is to say, it only needs to distinguish $|\varphi(m)\rangle$ and $|\varphi(m+M_b)\rangle$. However, Eve still needs to discriminate the M-ary quantum states. Due to the quantum noise's true randomness, when the noise is greater than the interval between adjacent quantum states, the information obtained by Eve will be confused.

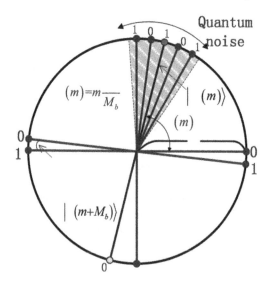

Figure 1.   Y-00 protocol encryption constellation.

126

3 SIMULATION SYSTEM MODEL

Figure 2 illustrates the PSK-QNRC simulation system model, which consists of four modules: transmitter, optical channel, decryption, and Bit Error Rate (BER) tester. In the transmitter, M_b-nary running key stream u and binary plaintext bit stream x generates an M-nary ($M = 2M_b$) ciphertext symbol stream off-line. A corresponding multi-digit electrical signal is generated by the ciphertext symbol stream in the signal source module, which is composed of a Pseudo-Random Binary Sequence (PRBS) generator and an Arbitrary Waveform Generator (AWG). Through a phase modulator, the electrical signal is modulated to the phase of the optical carrier generated by the laser. Then, the optical power is attenuated to the mesoscopic state by the attenuator, so that the quantum noise conceals the difference of neighboring states. Meanwhile, in order to adapt to long-distance fiber transmission, there is an amplifier at the transmitting end to amplify the small signal light to the classical power level.

The fiber channel is composed of two fibers and one amplifier. The second fiber is a Dispersion-Compensated Fiber (DCF) to compensate for the dispersion loss in the first fiber, that is, to satisfy $L_1D_1 + L_2D_2 = 0$. The amplifier is mainly used for power amplification to compensate for the transmission consumption of light power in the light fiber. When decrypted at the receiving end, Bob has the correct key, so the modulation state can be demodulated into the orthogonal ground state space for bit-by-bit selection, and the multi-ary electrical signal becomes a binary electrical signal. After decryption, the corresponding quantum state is:

$$|\varphi(m - u)\rangle = \left|\alpha e^{i(m-u)\frac{\pi}{M_b}}\right\rangle \tag{3}$$

According to the encryption mapping rule, because Alice and Bob share the key, so:

$$m - u = 0 \; or \; M_b, \, i.e \; (m - u)\frac{\pi}{M_b} = 0 \; or \; \pi \tag{4}$$

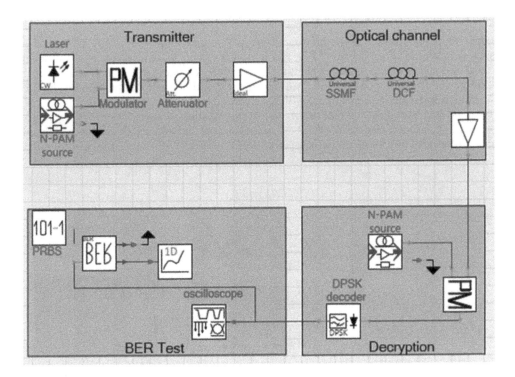

Figure 2. PSK-QNRC simulation system model.

We can see that, after decryption, the signal only carries phase 0 or π. Then, after the Differential Phase-Shift Keying (DPSK) transformation, it becomes an absolute code. Because Eve does not know the key, the decrypted signal still carries multi-phase information. Finally, the eye diagram is observed through an oscilloscope and the bit error rate is tested by the BER test module.

4 DISCUSSION

4.1 The impact of transmission distance and rate on transmission performance

Without loss of generality, taking 64PSK-QNRC as an example, the transmission distance and transmission rate of the fiber channel are changed. Considering the actual situation, the amplifier in the channel transmission is used to compensate for the transmission loss of the optical fiber, so when the transmission distance is changed, the gain of the amplifier is also changed. In the case of ensuring that other parameters in the system are unchanged, the change law of the bit error rate is as shown in Figure 3.

It can be seen from Figure 3 that as the transmission distance (L) increases, the BER increases continuously. Because of the noise in the fiber link and the amplifier, the signal-to-noise ratio continues to decline with the increase of L, so the performance of transmission becomes poorer. At the same time, as the transmission rate increases, the BER also rises. This is because as the rate increases, indicating that the number of bits transmitted per unit of time increases, so the BER also increases.

In this paper, we consider that when BER $P_e \leq 10^{-9}$, the system can be reliably transmitted. U-nder different transmission rates, when the transmission distance $d(2.5Gbit/s) = 228.6\,km$, $d(10Gbit/s) = 198.8\,km$, $d(15Gbit/s) = 189.6\,km$, $d(20Gbit/s) = 183.3\,km$, the BER $P_e = 10^{-9}$. It can be seen that as the transmission rate increases, the reliable transmission distance decreases. In order to ensure a high transmission quality, in the next discussion, we take the transmission rate as 10 Gbit/s and the transmission distance as 198 km.

4.2 The impact of average number of photons on transmission performance

Because the optical power is the photon energy per second, that is, $P = h\nu \cdot R \cdot N_{bit}$, where Planck's constant $h = 6.626 \times 10^{-34} J \cdot s$, the photon frequency ν corresponds to a wavelength of 1550 nm, and R is the bit rate. Therefore, $N_{bit} = P/h\nu R$. Figure 4 depicts the effect of the optical power after attenuation at the transmitting end and before amplification on the BER over 198 km with rate 10 Gbit/s under 64PSK-QNRC.

It can be seen from Figure 4 that the greater the attenuation, that is, the smaller the number of photons, the higher will be the BER. That is, the transmission reliability deteriorates. In order to ensure transmission performance, it is necessary to use appropriate optical

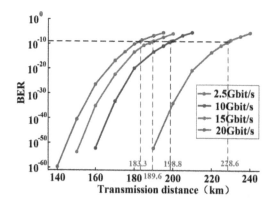

Figure 3. BER versus transmission distance with optical power after attenuation $P_{out} = -40dBm$ and $M = 64$.

128

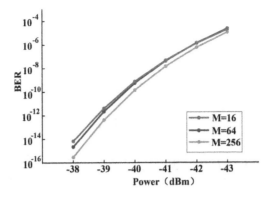

Figure 4.   BER versus optical power after attenuation P_{out} with L = 198 km and R = 10 Gbit/s.

amplification at the transmitting end. At the same time, we can see that the value of M may not affect the transmission performance to any extent, leading to our next research investigation.

4.3   The impact of M on transmission performance

It can be seen from Figure 5 that as M increases, the BER jitter does not change much, which means the change in M has little effect on transmission performance. In the next investigation, we will establish whether an increase in M can improve the security of the system. Therefore, Bob can improve security by increasing M without affecting transmission performance too much.

4.4   The impact of M on security performance

Eve can only obtain the multi-ary signal m_E because there is no key information. Under the cover of quantum noise, m_E is a random signal, so Eve cannot correctly demodulate the encrypted information. Figure 6 shows the eye diagrams for Eve when the hacking distance is 10 km under different values of M. As M increases, the degree of blurring of the eye diagram mounts up, indicating that the increase in M can improve the security of data transmission. This is because when M becomes larger, the phase difference of adjacent signals $\delta\varphi = 2\pi/M$ becomes smaller, and the number of signal levels masked by the quantum noise $NMS = \Delta\varphi/\delta\varphi$ rises correspondingly, so safety can be improved.

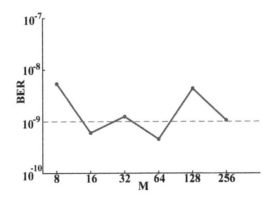

Figure 5.   BER versus M with $P_{out} = -40dBm$, L = 198 km and R = 10 Gbit/s.

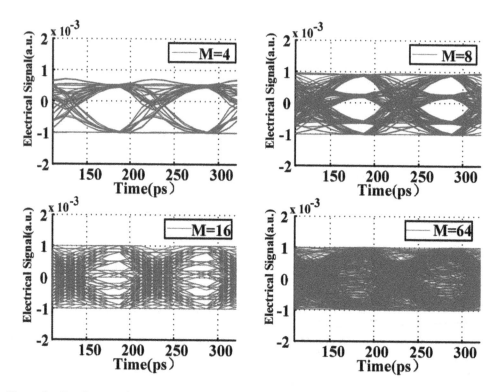

Figure 6. Eye diagrams for Eve under different values of M.

5 CONCLUSION

In conclusion, a PSK-QNRC anti-interception transmission simulation system model with single-channel data rates up to 10 Gb/s over 198 km fiber is proposed in this paper. To the best of our knowledge, it is the first research into anti-interception transmission using PSK-QNRC on the basis of standard commercial devices. We evaluate the effect of core parameters, such as transmission rate, transmission distance, average number of photons, and M, on the system transmission performance. Moreover, we analyze the impact of M on system security performance. Further quantitative calculations will be investigated in future work.

REFERENCES

Corndorf, E., Liang, C., Kanter, G.S., Kumar, P. & Yuen, H.P.-H. (2005). Quantum-noise randomized data encryption for wavelength-division-multiplexed-fiber optic network. *Physical Review A, 71*(6), 362–368.

Dai, W. (2012). *Continuous variable quantum cryptography communication experiment* (Master's thesis, Shanghai Jiao Tong University, Shanghai, China).

Jiao, H., Pu, T., Zheng, J., Zhou, H., Lu, L., Xiang, P., Zhao, J. & Wang, W. (2018). Semi-quantum noise randomized data encryption based on an amplified spontaneous emission light source. *Optics Express, 26*(9), 11587–11598.

Kanter, G.S., Reilly, D. & Smith, N. (2009). Practical physical-layer encryption: The marriage of optical noise with traditional cryptography. *IEEE Communications Magazine, 47*(11), 74–81.

Wang, S.X., Lipa, R.A., Reilly, D. & Kanter, G.S. (2013). Self-coherent differential phase detection for optical physical-layer secure communications. In *Optical Fiber Communication Conference/National Fiber Optic Engineers Conference 2013, OSA Technical Digest* (online), paper JW2A.41. Washington, DC: Optical Society of America.

Advances in Optoelectronic Technology and Industry Development – Jose & Ferreira (eds)
© *2020 Taylor & Francis Group, London, ISBN 978-0-367-24634-1*

Propagation characteristics of Super-Gaussian pulse in dispersion-decreasing fiber

Shengda Shi & Qiaofen Zhang
Key Laboratory of Precision Microelectronic Manufacturing Technology & Equipment of Ministry of Education, Guangdong University of Technology, Guangzhou, China

ABSTRACT: Based on the nonlinear Schrödinger equation (NLSE) and split-step Fourier method, the evolution equations of super-Gaussian pulses in Gaussian tapered dispersion-decreasing fiber (DDF) with anomalous group-velocity dispersion (GVD) are derived. The propagation characteristics of the super-Gaussian pulse and the influence of the initial chirp on the pulse propagation are both analyzed. In this paper, we discuss the transmission characteristics of super-Gaussian pulses with different initial chirp parameter C, that is C = 0, C = 2, and C = –2. The result shows that the initial chirp affects the transmission characteristics of pulse in time domain and frequency domain. When C = 0, the Super-Gaussian pulses has the best transmission waveform, spectral characteristics and chirp evolution characteristics.

Keywords: super-Gaussian pulses, dispersion-decreasing fiber, Initial chirp, time domain, frequency domain

1 INTRODUCTION

In modern fiber optical communication, signal transmission in single-mode fiber often causes the increase in transmission loss and the change in nonlinear effects due to an increase in transmission distance, and group-velocity dispersion and self-phase modulation imbalance due to signal distortion. A dispersion-decreasing fiber (Zhang. 2016)-(Zhang. 2011)-(Hirooka. 2004)-(Xu. 2016) with super-Gaussian pulses (Ali. 2017)-(Vinayagapriya. 2013)-(Zhong. 2016) longitudinal function is developed from the relationship between self-phase modulation and group-velocity dispersion. The second-order group-velocity dispersion in the dispersion-decreasing fiber is not a constant, but a variable that slowly decreases as the transmission distance increases, so the self-phase modulation caused by the loss is cancelled in the process to maintain the shape of the pulse. The amplitude of the pulse decreases with the loss of energy during transmission, but with appropriate gain the pulse does not change over long distances. Gaussian pulse is often used to analyze the transmission characteristics in the study of dispersion-decreasing fiber, However, in practical applications, the optical pulse emitted by the semiconductor laser is directly used as the transmitted optical pulse signal (Xu. 2018)-(Agrawal. 2013)-(Peng. 2016). Most of these signals are quasi super-Gaussian pulses pulses with steep front and rear edges. Super-Gaussian pulses are optical pulses with steep front and rear edges generated by directly modulated semiconductor lasers. When the optical pulse is transmitted in the dispersion-decreasing fiber with Super-Gaussian longitudinal function, the pulse width will change due to the combination of dispersion and nonlinear effect, which makes the signal easily distorted during signal transmission, and has a great impact on the accuracy and long-distance transmission of optical communication (Sunak. 2008).

Therefore, the nonlinear Schrödinger equation and the split-step Fourier method are used to explore the propagation characteristics of super-Gaussian pulses in the dispersion- decreasing fiber with super-Gaussian longitudinal function in this paper.

2 NONLINEAR SCHRÖDINGER EQUATION AND SPLIT-STEP FOURIER METHOD

The nonlinear Schrödinger equation is satisfied as follow when the optical pulse is transmitted in the fiber (Agrawal. 2013).

$$\mathrm{i}\frac{\partial A}{\partial Z} = -\frac{i}{2}\alpha A + \frac{1}{2}\beta_2\frac{\partial^2 A}{\partial T^2} - \gamma|A|^2 A \tag{1}$$

where A is the slowly varying amplitude of the pulse envelope; T is the time that moves with the group velocity of the pulse; Z is the transmission distance; β_2 is the second-order dispersion coefficient of the fiber; α is the loss coefficient; γ is the nonlinear coefficient. The generalization of super-Gaussian pulses is as follows:

$$A(0, T) = \exp\left[\frac{1 + iC}{2}\left(\frac{T}{T_0}\right)^{2m}\right] \tag{2}$$

where C is the chirp parameter; T is the time that moves with the group velocity of the pulse; T_0 is the initial input pulse; m is the steepness of the front and rear edges. In order to apply the split-step Fourier method, Equation (1) is written in the following form:

$$\frac{\partial A}{\partial Z} = (\hat{D} + N)A \tag{3}$$

where $\hat{D}$ is the difference operator, which represents the dispersion and absorption of the linear medium, and N is the nonlinear operator, which determines the influence of the nonlinear effect of the fiber during the transmission. Here we can obtain the following expression:

$$\hat{D} = \frac{i}{2}\beta_2\frac{\partial^2 A}{\partial T^2} \tag{4}$$

$$N = i\gamma|A|^2 \tag{5}$$

To study the influence of the dispersion and nonlinear effects of the fiber on the pulse transmission, it is ideal to ignore the loss of the fiber to satisfy the following equation:

$$i\frac{\partial A}{\partial Z} = \frac{1}{2}\beta_2\frac{\partial^2 A}{\partial T^2} - \gamma|\mathrm{A}|^2 \tag{6}$$

γ is a nonlinear coefficient, β_2 is a variable, $\beta_2 = \beta_2(0)\mathrm{P}(Z)$, where P(Z) is the longitudinal variation of dispersion. In order to analyze the dispersion coefficient characteristics of the dispersion decreasing fiber, we take the Fourier transform on the dispersion part of A(Z, T):

$$A(Z, w) = A((0, \omega)\exp\left[\frac{i}{2}\omega^2 h(Z)\right] \tag{7}$$

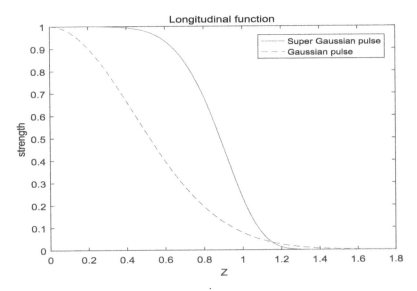

Figure 1. Group dispersion longitudinal variation.

where h(z) = $\beta_2(0) \int_0^z p(z)dz$. The dispersion longitudinal change of dispersion-decreasing fiber is in the form of super-gaussian function, the expression can be written as:

$$P(Z) = exp[-(Z^{2m}/L^{2m})lnk] \qquad (8)$$

The image of the longitudinal dispersion of the group dispersion when the dispersion is gradually decreased by super Gaussian and Gaussian is shown in Figure 1.

It can be seen from Figure 1 that after 1.18 km of transmission, the slope of any super-Gaussian dispersion variation is larger than that of Gaussian dispersion, and the overall mode of the curve is steeper. For different group velocity longitudinal variable functions, the ratio of compression after optical pulse transmission is different.

3 THE EFFECT OF INITIAL CHIRP ON SUPER-GAUSSIAN PULSES

3.1 *Time domain analysis of pulse transmission*

The shape of the pulse does not change when the super-Gaussian pulse is transmitted in the dispersion-decreasing fiber with a super-Gaussian longitudinal function, but the broadening factor also changes as the transmission distance increases. Ideally, only when the dispersion effect of the fiber acts, according to the comparative analysis of the three cases, It can be seen that the initial chirp has an influence on the broadening factor of the super-Gaussian pulse in the transmission process of the dispersion-decreasing fiber, so as to analyze the transmission characteristics of the super-Gaussian pulse.

It can be seen from the calculation that when c = 0, the characteristics of the broadening factor of the super-Gaussian pulse without initial chirp in the dispersion-decreasing fiber with super Gaussian longitudinal function as a function of distance are shown in Figure 2.

In the absence of frequency chirp, the broadening of the pulse is only determined by the group velocity dispersion and the nonlinear effect. It can be seen from Figure 2 (A) that the broadening factor is firstly reduced to 1 and then increased rapidly to a constant 1.19.

The broadening factor of the super-Gaussian pulse is shown in Figure 2 (B) when the initial chirp c = –2. As shown in Figure 2 (B), the curve of the broadening factor is determined by the initial chirp and the initial pulse shape. When the transmission distance is 0.68 km, the

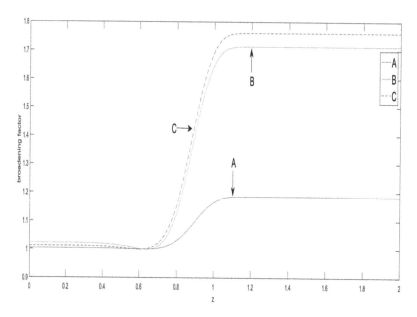

Figure 2. Curve of the broadening factor as a function of transmission distance at c = 0, c = –2, c = 2.

spreading factor is reduced to the lowest value which is equal to 1. As the fiber length increases, the spreading factor increases to a constant 1.77. This is because when $\beta_2 c > 0$, the chirp generated by group velocity dispersion is the same as the initial positive chirp, causes the total chirp to increase. Therefore, when the transmission distance is the same, the broadening factor is larger when c = –2.

When c = 2, the relationship between the broadening factor of the super-Gaussian pulse transmitted in the dispersion- decreasing fiber with super-Gaussian longitudinal function and distance is shown in Figure 2 (C). When the initial pulse is positively chirp, the super-Gaussian pulse exhibits initial narrowing in the dispersion-decreasing fiber and the spreading factor reaches to a lowest value 1 at z = 0.68 km. As the transmission distance increases, the spreading factor increases to 1.73 and remain constant. This can be interpreted by the chirp effect. When the pulse is initially chirped and the condition $\beta_2 c < 0$ is satisfied, the dispersion-induced chirp is in opposite direction to that of the initial chirp. As a result, the net chirp is reduced, leading to pulse narrowing. The minimum pulse width occurs at a point at which the two chirps cancel each other. With a further increase in the propagation distance, the dispersion-induced chirp starts to dominate over the initial chirp, and the pulse begins to broaden. Due to the reduction in total chirp, the broadening factor changes slowly with increasing distance. The degree of pulse broadening depends on the steepness of the initial chirp and super-Gaussian pulse. Under the effect of dispersion and nonlinearity, the propagation characteristics of the super-Gaussian pulse in the dispersion-decreasing fiber with super-Gaussian longitudinal function are shown in Figure 3A.

Figure 3A shows that as the transmission distance increases, peak occurred when transmitting 0.31 km, and a multi-peak structure appears, the energy is mainly stored in the main peak. This is mainly due to the modulation instability generated in the anomalous dispersion region of the fiber, resulting in a pulse-generating multimodal structure. A lot of noise began to appear when the transmission distance increased to 0.45 km, and the intensity of the pulse is about 100 W. This occurs first because the different frequency components in the pulse propagate in the anomalous dispersion region of the fiber at different group velocities. The modulation instability caused by the imbalance between dispersion and nonlinearity affects the shape and distribution of the pulse. Secondly, because the input pulse is a super-Gaussian pulse, the super-Gaussian pulse has a steep front and rear edge, which is prone to pulse broadening and signal distortion.

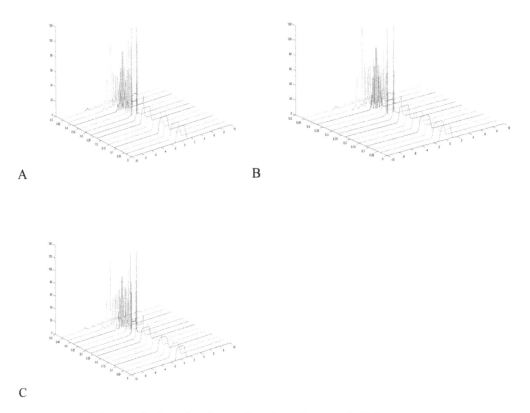

A   B

C

Figure 3.   Variation of pulse intensity of super-Gaussian pulse at c = 0, –2, 2.

In the case that the pulse has an initial chirp, the degree of pulse broadening depends on the sign of $\beta_2 c$. When the initial chirp c = –2, $\beta_2 c > 0$, the sign of the chirp caused by the dispersion is the same as the initial, making the total chirp larger, and the broadening factor becomes larger in the process of transmitting the same distance, there by further increasing The intensity of the pulse broadens. As shown in Figure 3B. If the incident super-Gaussian pulse has a negative chirp (c < 0). During the pulse transmission, the highest peak appeared when the signal was transmitted at 0.31 km, and a similar multimodal structure also appeared. The waveform changes similarly to the waveform change of the super-Gaussian pulse without initial chirp. However, the pulse intensity is different. This is because the initial pulse is a negative pulse $\beta_2 c > 0$, so that the effect caused by dispersion is dominant. Conversely, if the incident super-Gaussian pulse has a positive chirp (c>0), the pulse waveform change will be gentler than that without the initial chirp, as shown in Figure 3C, the pulse intensity of the super-Gaussian pulse changes. When $\beta_2 c < 0$, the pulse is narrowed and then broadened during transmission. During the pulse transmission, the highest peak appeared when the signal was transmitted at 0.31 km, and a similar multimodal structure also appeared. After the transmission distance of 0.45 km, noise starts to appear, and the pulse intensity is about 130 W. This phenomenon occurs because when the incident super-gaussian pulse with a positive initial chirp and the frequency chirp caused by fiber dispersion work together, the initial negative chirp effect weakens the effect caused by the frequency chirp caused by fiber dispersion, so the results shown in Figure 3C appear.

To better observe the change of the transmission characteristics of the super-Gaussian pulse during transmission, the curve at z = 0.31km is selected for observation. By comparing the super Gaussian pulse with initial chirp and the super-Gaussian pulse without initial chirp, the transmission characteristics of the super-Gaussian pulse are analyzed. Figure 4 shows the characteristic curve of super-Gaussian pulse transmission.

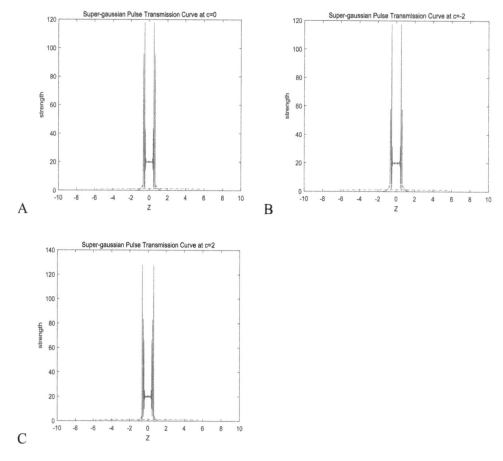

Figure 4.   c = 0, c = 2, c = –2 for the transmission characteristics of super-Gaussian pulses.

The case where the pulse energy is the highest in three cases is selected for analysis. When there is no initial chirp, the pulse shows a multi-peak structure, and the main energy is concentrated on the main peak, the main peak disappears when the energy reaches 22 W, and the highest peak is on the inner side of the multi-peak structure. When the energy reaches 98 W, the outermost slant disappear. When the energy reaches 118 W or so, the pulse reaches its maximum value. When $\beta_2 c > 0$, the pulse has a multi-peak structure, the energy is mainly concentrated on the main peak, and the highest peak is on the inner side of the multi-peak structure. When the energy reaches 58 W, the outermost slant disappears, and the main peak disappeared when the energy reaches 22 W or so, the pulse reached its strongest when the energy reached 118 W. The reason is that the chirp caused by dispersion has the same sign as the initial chirp, and the superposition of the two interaction with weaker nonlinear effects. When $\beta_2 c < 0$, the pulse also has a multi-peak structure. The energy is mainly concentrated on the main peak. The highest peak appears on the outermost side of the multi-peak structure. When the energy reaches 22 W, the main peak disappears. When the energy reaches 135 W, the pulse reached its maximum. The reason for this is that the chirp caused by dispersion is different from the initial chirp, and the two compensate each other, so that the effect of the chirp caused by dispersion is weaker than that produced when there is no chirp, and the dispersion effect cannot balance the nonlinear effect, so that the pulse is mainly affected by the nonlinear effect.

3.2 *Frequency domain analysis of pulse transmission*

The spectrum of the pulse depends not only on the shape of the pulse, but also on the initial chirp of the pulse. Reflecting changes in the spectrum based on changes in initial chirp, discusses the effect of chirp on the bandwidth of the pulse spectrum. Figure 2.8 shows the super-Gaussian pulse transmission spectrum.

From Figure 5, the variation of the super-Gaussian pulse transmission spectrum can be seen. The multi-peak structure appears in the spectrum after 0.13 km of transmission because the spectral characteristics are affected by the phase modulation, and the same size of chirp appears at two different points, which have the same instantaneous frequency. These two points represent two waves of the same frequency but different phases, and additive or destructive interference occurs according to their relative phase differences, forming a multi-peak structure of the pulse spectrum. The spectrum of the Gaussian pulse with chirp is wider than that of the super-Gaussian pulse without the initial chirp, the spectrum at $c = 2$ is similar to the spectrum without initial chirp and the spectrum at $c = -2$ is quite different from the spectrum without initial chirp. In order to better analyze this structure, the waveform of 0.1 km transmission is selected for observation and analysis. As shown in Figure 6.

It can be seen from Fig 2.9 whether there is a difference in the super-Gaussian pulse spectrum at the initial chirp, and the pulse at $c = 2$ makes the pulse appear smoother and the pulse broadening is larger with respect to $c = 0$. This is because the initial chirp at $c = 2$ is the same as the chirp generated by the self-phase modulation, and the initial chirp is superimposed with the chirp generated by the self-phase modulation, resulting in an enhancement of the oscillation structure, resulting in pulse broadening. However, when $c = -2$, the waveforms of the first two cases are significantly different. The pulse block exhibits a multi-peak structure and the intensity

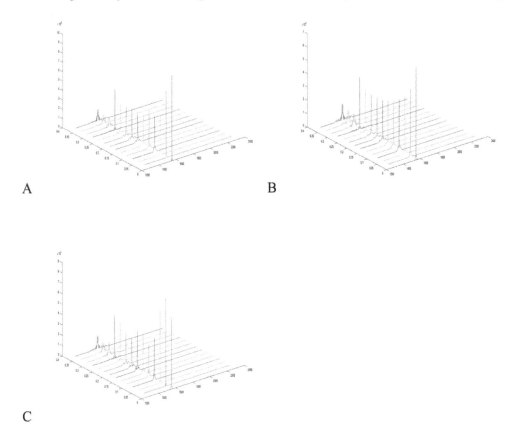

A B

C

Figure 5.   Spectral characteristics of super-Gaussian pulses at $c = 0$, $c = 2$ and $c = -2$.

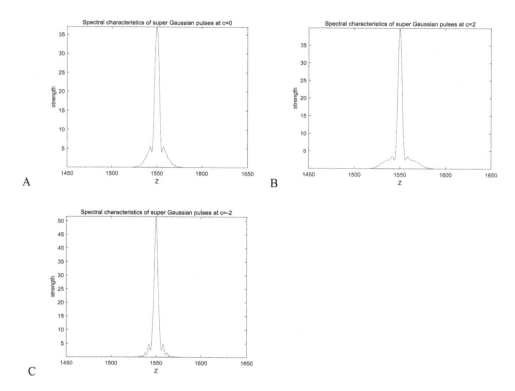

Figure 6.   Partial spectrum of super Gaussian pulses at c = 0, c = 2 and c = –2.

of the pulse is relatively low relative to the pulse without initial enthalpy. The main reason for this difference is that when the negative chirp of super-Gaussian pulse and the positive chirp caused by self-phase modulation work together, the spectrum is narrowed. The formation of the pulse block is mainly due to the fact that the red-shifted portion of the pulse in the anomalous dispersion region cannot keep up with the fast-moving blue shift, causing the energy of the tail to diverge, thereby creating a pedestal.

All three cases are assumed that the loss of the optical fiber is negligible. The three spectra are different, and the steep front and back edges of the super-Gaussian pulse will distort the shape of the pulse greatly with the increase of the distance. Usually within a certain range, the spectrum exhibits an oscillating structure over the entire frequency range, and the spectrum is a multi-peak structure with the maximum intensity of the outermost peak, but most of the energy is still retained in the central peak. This is because the $|T|<T0$ is a super-Gaussian pulse with almost uniform light intensity. The reason why the spectrum is continuously broadened during transmission is because the new frequency components caused by self-phase modulation are continuously generated, and secondly. This is because modulation instability is likely to occur in the anomalous dispersion region of the fiber, and modulation instability affects the pulse distribution.

4   CONCLUSION

Based on the nonlinear Schrödinger equation and split-step Fourier method, the transmission characteristics of a super-Gaussian pulse in dispersion-decreasing fiber with super-Gaussian longitudinal function and the influence of initial chirp on transmission parameters are derived, it is concluded that:

(1) The super-Gaussian pulse has steeper front and rear edges compared with the Gaussian pulse, and the dispersion loss in the transmission process is larger than the initial dispersion loss of other pulses, which makes the pulse broaden faster.

(2) The frequency chirp is generated during the super-Gaussian pulse transmission. When the negative initial chirp is carried, the variation of the pulse transmission waveform is similar to the change of the super-Gaussian pulse without initial chirp. When carrying a positive initial chirp for transmission, the pulse has a process of first narrowing and then broadening, and the waveform change is more gentle than that of the incident of the super-Gaussian pulse without the initial chirp. In the transmission process, the pulse transmission of the three cases shows a multi-peak structure after a period of time, but the energy of the pulse is deviated, and as the transmission distance increases, the waveform begins to distort and finally replaced by noise.

(3) By observing the change of the super-Gaussian pulse spectrum, the spectral characteristics of the super-Gaussian pulse with positive initial chirp are more obvious than the spectral oscillation structure of the super-Gaussian pulse without initial chirp, and the pulse broadening is larger. The spectral characteristics when carrying a negative initial chirp are compared to the spectrum of a super-Gaussian pulse without initial chirping, and a multi-peak structure appears in the pulse holder, pulse is compressed.

ACKNOWLEDGMENTS

This work was supported by National Natural Science Foundation of China (No. 61705045, No. U1601202, No.51675106), Guangdong Provincial Science and Technology Research Project ((No.2015B010104008, No.2016A030308016, No.17ZK0091)), Foundation of Guangdong Province Science and Technology (No. 2017A090905047), Special and Technology Enterprises of Provincial Science and Technology Enterprises of Small and Medium Size (No. 2016A010119143), Foundation of Guangdong Province Science and Technology (No. 201604010011).

REFERENCES

F. Xu, J. Yuan, C. Mei, F. Li, Z. Kang, B. Yan, X. Zhou, Q. Wu, K. Wang, X. Sang, C. Yu, G. Farrell, 2018. Mid-Infrared Self-Similar Pulse Compression in a Tapered Tellurite Photonic Crystal Fiber and Its Application in Supercontinuum Generation, J. Light. Technol. doi:10.1109/JLT.2018.2839520.

G. Agrawal, G.P. Agrawal, 2013. Nonlinear Fiber Optics. doi:10.1016/B978-0-12-397023-7.00005-X.

H.R.D. Sunak, 2008. Optical fiber communications, Proc. IEEE. doi:10.1109/proc.1985.13332.

Q.F. Zhang, Y.H. Deng, 2016. Influence of gain coefficient on the self-similar pulses propagation in a dispersion-decreasing fiber, Optik (Stuttg). doi:10.1016/j.ijleo.2016.02.060.

Q. Zhang, J. Gao, 2011. Generation of excellent self-similar pulses in a dispersion-decreasing fiber, Optik (Stuttg). 122 1753–1756. doi:10.1016/j.ijleo.2010.10.037.

R. Ali, M.Y. Hamza, 2017. Propagation behavior of super-Gaussian pulse in dispersive and nonlinear regimes of optical communication systems, ICET 2016-2016 Int. Conf. Emerg. Technol. doi:10.1109/ICET.2016.7813264.

S. Vinayagapriya, A. Sivasubramanian, PMD induced broadening on propagation of Chirped Super Gaussian pulse in single mode optical fiber, Int. Conf. Signal Process. Image Process. Pattern Recognit. 2013, ICSIPR 2013. doi:10.1109/ICSIPR.2013.6497947.

T. Hirooka, M. Nakazawa, 2004. Parabolic pulse generation by use of a dispersion-decreasing fiber with normal group-velocity dispersion, Opt. Lett. doi:10.1364/ol.29.000498.

X. Zhong, D. Liu, J. Sheng, 2016. Long-term nonlinear propagation and damped oscillation behaviors of Gaussian and super-Gaussian pulses in optical fibers, Optik (Stuttg). doi:10.1016/j.ijleo.2016.04.118.

Y. Peng, S. Wang, R. Li, H. Li, H. Cheng, M. Chen, S. Liu, 2016. Luminous efficacy enhancement of ultraviolet-excited white light-emitting diodes through multilayered phosphor-in-glass, Appl. Opt. doi:10.1364/ao.55.004933.

Y. Xu, H. Ye, D. Ling, G. Zhang, 2016. Effect of initial chirp on supercontinuum generation in dispersion decreasing fibers, Optik (Stuttg). doi:10.1016/j.ijleo.2015.10.205.

Advances in Optoelectronic Technology and Industry Development – Jose & Ferreira (eds)
© 2020 Taylor & Francis Group, London, ISBN 978-0-367-24634-1

Time-delay measurement of optical-fiber link based on time-frequency simultaneous transmission method

J.C. Guo, L. Lu, H. Wei & X.Y. Zhao
Institute of Communication Engineering, Army Engineering University of the People's Liberation Army, Nanjing, Jiangsu, China

ABSTRACT: In order to meet the requirements of time-delay measurement in optical-fiber links, we propose a scheme of joint transfer of frequency and pulse-per-second time signals at the same wavelength. The scheme combines the coarse results of a pulse-per-second time-counting method and the fine results from a frequency signal to achieve high-precision and large-range measurement of the true delay of a fiber link. We build up an experimental system to measure the absolute delay of the signal within 25 km optical fibers under temperature variation. The experimental results show that the method can effectively combine the large-range advantage of the one pulse-per-second counting method with the high-resolution advantage of the phase-measurement method.

1 INTRODUCTION

High-precision, large-range measurement of time delay in optical fibers has important applications in systems such as time-frequency transmission, radio over fiber, and distributed stations (Tashiro et al., 2012; Jiang et al., 2015). At present, time-delay measurement methods based on optical-fiber links include two categories: one is based on high-speed counting, Vernier range, delay line interpolation method, pulse time-of-flight measurement method (Yu et al., 2013; Fujieda et al., 2009; Ebenhag et al., 2013); the measurement range of these methods can be up to the level of seconds, but the resolution of mature measuring instruments or chips based on such techniques is generally tens of picoseconds, Some recently reported delay-measuring tools can reach the order of picoseconds, which can be considered as a "coarse value" measurement method. The other scheme is coherent measurement, which can be considered as a "fine value" measurement method by using optical RF and optical combs (Hanssen et al., 2011; Zhou et al., 2017) to compare the phase of the round-trip signal; the resolution of this delay measurement can even reach the level of femtoseconds. With phase ambiguity, the measured range is usually limited to tens of kilometers, and long-distance measurement of a large dynamic range cannot be realized. Methods of zero-phase fixation and modulation phase shift combined with coarse measurement and fine measurement explore the simultaneous implementation of large-range, high-precision measurement demand, but also have the problem of insufficient resolution, high uncertainty, complexity, and unstable measurement.

In this context, we propose a scheme of joint transfer of frequency and pulse-per-second time signals based on the same wavelength. The scheme combines the coarse results of the pulse-per-second time-counting method and the fine results from a frequency signal comparison to achieve high-precision and large dynamic measurement of the true delay of a fiber link. We build up experimental system to measure the absolute delay of the signal within 25 km optical fibers in severe temperature change. The experimental results show that the method can effectively combine the large-range advantage of the one pulse-per-second counting method with the high-resolution advantage of the phase-measurement method.

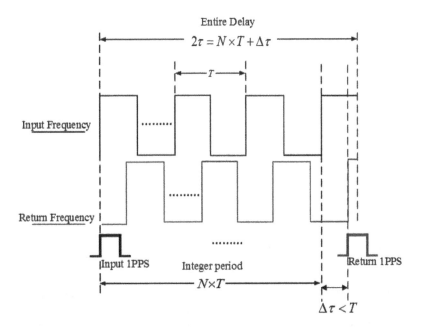

Figure 1. Principle of time-delay measurement.

2 PRINCIPLE OF TIME-DELAY MEASUREMENT OF OPTICAL FIBER

The core of the method in "time-frequency simultaneous transmission" is that the transmitting module modulates the time pulse signal (1 Pulse Per Second, 1PPS) and the frequency signal onto the same optical carrier, and the two signals are simultaneously transmitted in the optical fiber to ensure that their delays are identical. The principle is illustrated in Figure 1. First, a homologous in-phase 1PPS signal ('Input 1PPS') and frequency signal ('Input Frequency') are generated by the same atomic clock, and the two signals are modulated with the same laser and transmitted into the same fiber. If the period of the atomic clock output frequency signal is T, then the time interval between two rising or falling edges of the adjacent frequency signal will be strictly equal to T. Thus, any link time interval can be divided into integer multiples and less-than-one-period parts. Assuming that the one-way delay is τ, where less than one integer period part is $\Delta\tau$, then the total time delay 2τ is:

$$2\tau = N \times T + \Delta\tau \tag{1}$$

The total delay 2τ includes two parts, of coarse value $N \times T$ and fine value $\Delta\tau$; this is shown in Figure 1. The integer time delay of the first part is measured by a 1PPS-triggered time counter. The second part is a delay of less than one period, which is obtained by comparing the phase difference between the local frequency signal ('Input Frequency') and the returned frequency signal ('Return Frequency') at the phase comparator. Taking a frequency signal of 100 MHz as an example, because the period is 10 ns, the coarse value can be taken to be of the order of 10 ns, and the fine value within 10 ns is measured by the round-trip 100 MHz signal phase comparison. Adding the coarse measured value to the fine measured value gives the measured Fiber-link Time Delay (FTD) value.

The time-frequency simultaneous transmission measurement scheme is shown in Figure 2. The system consists of five parts: a time-frequency transmitting module, optical-fiber link, receiving module, coarse measurement part, and fine measurement part. First, a clock generator ('CLK') outputs three 100 MHz frequency signals referenced by a local rubidium atomic

141

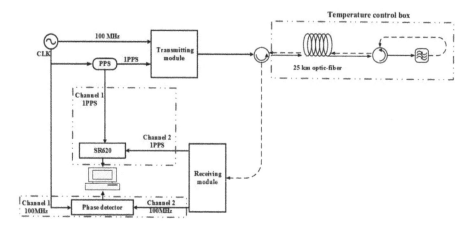

Figure 2.   Scheme of time-delay measurement.

clock. One of the 100 MHz signals is connected to the first channel ('Channel 1') of the phase detector as the local frequency comparison signal; another 100 MHz signal is connected to the PPS module to generate the 1PPS signal. One 1PPS signal is connected to the first channel ('Channel 1') of the SR620 time-interval measuring instrument as the start signal, another 1PPS signal is combined with the third 100 MHz signal. This combined 1PPS and 100 MHz signal is modulated by the laser onto a wavelength of 1545.3 nm. The optical transmission signal is connected to the G.652 single-mode fiber through a circulator. The end of the fiber is connected to another circulator, and the filtered return optical signal is connected to the second input port of the circulator and comes back through the original fiber. The return-ing optical signal is detected by the PIN (photo-detector). The electrical signal from the PIN is divided into two parts, which are filtered and amplified independently to recover the 100 MHz and 1PPS signals to satisfy the measurement requirement. The recovered 1PPS signal is con-nected to the second channel ('Channel 2') of the SR620, so that the integer part of the time delay is measured. The recovered 100 MHz frequency signal is connected to the second chan-nel ('Channel 2') of the phase comparator, so that the fine value of the time delay is measured. The sum of the coarse value and fine value is the FTD value. The FTD value is linear super-imposed by the coarse value and fine value.

3   EXPERIMENTAL RESULTS

3.1   *Optical-fiber link time-delay measurement under stable temperature*

The experiment first measures the initial time delay of the system and then measures the 25 km fiber-link time delay at a constant temperature.

In a stable indoor environment of 24–26°C, the time delays of the system are shown in Figure 3, and the white dotted lines indicate the results after Kalman filtering. In Figure 3a, the coarse value of the system fluctuates within the range of 88.6–88.7 ns. According to the measurement principle, the coarse measurement result is 80 ns. Fine measurement results in Figure 3b show the background delay range is 8.790–8.795 ns, with a mean time delay of 8.79415 ns. Therefore, the FTD value is 88.79415 ns under the initial system. We then replace the short fiber with a 25 km fiber. The coarse value for the 25 km fiber-link system is shown in Figure 3d. The average value is between 247975.1 and 247978.3 ns, and the coarse value for the short optical fiber was 80 ns, so the 25 km coarse measurement result is 247890 ns. In the fine measurement results, due to the slow fluctuation of the optical-fiber delay, the 25 km link fine measurement delay is within 50 ps, as shown in Figure 3e before the start of temperature control.

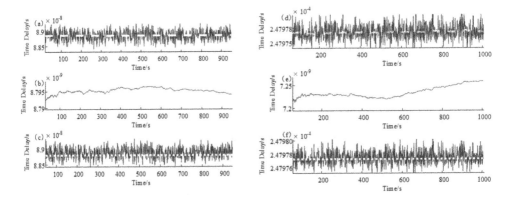

Figure 3. Time delay of system at constant temperature: (a) 1PPS+100 MHz coarse value; (b) 1PPS +100 MHz fine value; (c) 1PPS+100 MHz FTD value; (d) 25 km 1PPS+100 MHz coarse value; (e) 25 km 1PPS+100 MHz fine value; (f) 25 km 1PPS+100 MHz FTD value.

3.2 25km optical-fiber link time-delay measurement under temperature variation

The 25 km optical fiber is placed in the temperature-control box, and the temperature change in the box is used to simulate temperature change in the actual environment. In order to clearly reflect the influence of the link temperature change on the delay fluctuation, the temperature of the box is set to rise from 25 to 35°C under continuous operation for 5000 s, then the door is opened and the temperature inside the box is lowered to environmental temperature; because the 25 km fiber in the experiment is the whole disk fiber, the constant temperature for 5000 s ensures that the fiber temperature continues to rise. When the ambient temperature changes by 10°C and the fiber delay variation coefficient is 30 ps (km•°C), the theoretical delay variation is 15 ns. The unidirectional propagation delay measured in the experiment is changed by about 15.8 ns in Figure 4a, which is consistent with the theory and the reports in the related literature.

Figure 4 shows the coarse and fine measurement results of the temperature rise in the box for 5000 s. The results show that the coarse value of the system shown in Figure 4b is changed from that of the original 25 km stable environment of 247978 ns to 247993 ns. The fine value for the optical fiber, in Figure 4b, changed more than two periods. It can be seen changing from 7.63253 to 8.24756 ns during the temperature rise over 25000 s, which is a significant

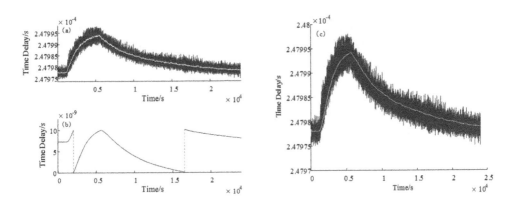

Figure 4. Coarse and fine values of 25 km fiber link under temperature variation: (a) 25 km, temperature variation, 1PPS+100 MHz coarse value; (b) 25 km, temperature variation, 1PPS+100 MHz fine value; (c) 25 km, temperature variation, 1PPS+100 MHz FTD value.

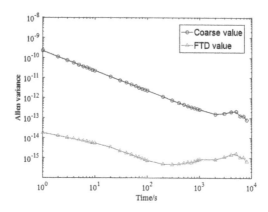

Figure 5. Allen variance.

improvement over the coarse value. The two result values spliced together are shown in Figure 4c. The FTD value indicated by the white line fits well with the rough measurement result indicated by the black line, and the fluctuation of the FTD value is significantly smaller than that of the coarse value. Subtracting the initial system time delay of 88.79415 ns, the FTD value is in the range 247977.63253–247978.24756 ns.

The system Allen variance is shown in Figure 5. It can be seen from the figure that the second Allen variance of the coarse measurement is 3.54×10^{-10}, and the second Allen variance of the FTD value is 2.72×10^{-14}, so the second Allen variance is greatly improved under the FTD measurement. According to the correspondence between the second Allen variance and the measurement accuracy, the accuracy is about 0.03 ps. It is obvious that the FTD value has better accuracy than the coarse value and this verifies the feasibility of the scheme under long-distance temperature changes.

4 CONCLUSION

In conclusion, we proposed a method of coarse and fine combination measurement in time-frequency co-transmission for optical-fiber time delay. The measurement scheme was designed and a platform built for experimentation. The time-delay measurement of the 25 km fiber link in the laboratory environment was completed. The absolute delay and delay fluctuation measurement of 0.03 ps resolution is realized under the condition of 15 ns delay fluctuation. The range of the method is determined by the coarse measurement and the resolution is determined by a finer measurement than the phase method, so that the advantages of the two methods can be effectively combined to achieve high-precision, large-range measurement of optical-fiber time delays.

REFERENCES

Ebenhag, S.C., Hedekvist, P.O. & Jaldehag, K. (2013). Two-color one-way frequency transfer in a metropolitan optical fiber data network. *NCSLI Measure, 8*(2), 52–61.

Fujieda, M., Kumagai, M., Gotoh, T., Hosokawa, M. et al. (2009). Ultrastable frequency dissemination via optical fiber at NICT. *IEEE Transactions on Instrumentation & Measurement, 58*(4), 1223–1228.

Hanssen, J.L, Crane, S.G. & Ekstrom, C.R. (2011). One-way temperature compensated fiber link. *Proc. Joint Conf. IEEE IFCS*, 1–5.

Jiang, J.Z., Dai, Y.T., Zhang, A.X., Yin, F.F. Li, J.Q., Xu, K., Lv, Q., Ren, T.P., Tang, G.S. et al. (2015). Precise time delay sensing and stable frequency dissemination on arbitrary intermediate point along fiber-optic loop link with RF phase locking assistance. *IEEE Journal on Photonics, 7*(2), 1–9.

Tashiro, T., Miyamoto, K., Lwakuni, T. Hara,K., Fukada, Y., Kani, J., Yoshimoto, N., Iwatsuki, K., Higashino, T., Tsukamoto, K., Komaki, S. et al. (2012). 40 km fiber transmission of time domain multiplexed MIMO RF signals for ROF-DAS over WDM-PON. *Conference on Optical Fiber Communication*, 1–3.

Yu, L.Q., Lu, L., Wang, R., Wu, C.X., Zhu, Y., Zhang, B.F., et al. (2013). Analysis of the Sagnac effect and its influence on the accuracy of the optical fiber time transfer system. *Acta Optica Sinica, 33*(3), 0306003.

Zhou, W.F., Shi, J.K., Ji, R.Y., Li, X. & Liu, Y. (2017). High-precision distance measurement using femtosecond laser frequency comb. *Chinese Journal of Scientific Instrument, 38*(08), 1859–1868.

Advances in Optoelectronic Technology and Industry Development – Jose & Ferreira (eds)
© 2020 Taylor & Francis Group, London, ISBN 978-0-367-24634-1

SNR uniformity optimization for LEDs ring alignment in visible light communications

Fang Li, Jia Xiao, Xiaojun Li, Jun Xia, Qiaoting Zhou & Guosheng Hu
Communications and Information Department, Shanghai Technical Institute of Electronics & Information, Shanghai, P. R. China

ABSTRACT: In this paper, we proposed an optimization for Light-Emitting Diode (LED) ring alignment, to obtain optimal Signal-to-Noise Ratio (SNR) uniformity. Compared with the non-optimized situation, results show that the variance of SNR can be reduced from 8.4 dB to 2.1 dB. Moreover, we also investigated the optimization for LED ring-corner alignment; it demonstrated superior overall performance of SNR and illuminance, since the performance for corners and edges improved.

Keywords: LED ring alignment, uniformity, SNR, illuminance, SAHP, visible light communication

1 INTRODUCTION

Since Light-Emitting Diodes (LEDs) are widely used in indoor lighting, visible lighting communication employing LEDs is arousing wide interests nowadays (Komine & Nakagawa, 2004; Borogovac et al., 2010; Wu et al., 2011; Tanaka et al., 2003). Due to the advantages of high security, unlicensed spectrum and high modulation rate, LEDs can provide data transition by high-rate switching on and off while lighting (Boucouvalas et al., 2015; Komine et al., 2001). Most research has focused on designing the modulation scheme (Kwon, 2010; Kim & Jung, 2011; Rajagopal et al., 2012) and employing the LED dimming control (Sung et al., 2014; Noh et al., 2015).

Uniformity of lighting and communications are important in Visible Light Communication (VLC), because they guarantee uniform optical power in all positions, resulting in better coverage. A trial-and-error approach to distributing optical power evenly was proposed for circular LED alignment (Wang et al., 2012). An evolutionary algorithm was proposed to modify received power fluctuation (Ding & Ji, 2012). A genetic algorithm to optimize the reflected indices of the concentrator to obtain uniform distribution was reported (Higgins & Green, 2015). A heuristic power distribution method was proposed for random LED alignment to achieve uniform lighting (Varma et al., 2017).

Moreover, the Semi-Angle At Half Power (SAHP) of LEDs is the parameter that denotes the divergence of the LED emission. It determines the received power and hence influences the uniformity performance of illuminance and Signal-to-Noise Ratio (SNR) distribution. The uniformity performances of groups of LED alignments are investigated in Li et al. (2015), and Li et al. (2016), which demonstrate that the SAHP optimizations are capable of increasing the lighting and communication distributions.

In this paper, we proposed an optimization of LED alignment parameters and LED SAHP, for LED ring and ring-corner alignments. To obtain SNR uniformity, we chose the Q-factor of the SNR, Q_{SNR}, (Li et al., 2016) as the optimization goal, with the constraint of minimum communications SNR requirement. For LED ring alignment, the optimal SNR variance can be reduced by over 6 dB, and the illuminance distribution can also be improved. To further increase the performance at the corner area, we located LEDs at the ceiling corner, and implemented the optimization, involving parameters of the corner LED's SAHP and positions, to be determined jointly. Results show that, compared to the LED ring alignment, the edge and corner performance can both be improved. Hence the optimization produces overall superior performance for the LED ring-corner alignment.

In the next section, the models of lighting and communication models are investigated. Then, in Section 3, we demonstrate the optimization process and results, both for LED ring alignment and LED ring-corner alignment. Conclusions are given in Section 4.

2 THEORETICAL MODELS

2.1 Illuminance models

The illuminance of a place by an LED is given as (Komine & Nakagawa, 2004):

$$E = I(0)\cos^m(\theta)\cos(\psi)/d^2, \tag{1}$$

where $I(0)$ is the center luminous intensity of an LED, θ and ψ are the angle of irradiance and incidence, respectively, and d is the distance between the LED and receiver. The order of Lambertian emission, m, is given by the SAHP of an LED, $\theta_{1/2}$, as (Komine & Nakagawa, 2004):

$$m = \ln2/\ln(\cos(\theta_{1/2})) \tag{2}$$

The total illuminance by several LED sources are the sum of individual illuminances.

2.2 Communication models

The received power of a given location is denoted as (Komine & Nakagawa, 2004):

$$P_r = \sum_{i=1}^{LEDs} h_i(t) \times P_t, \tag{3}$$

where P_t is the transmitted optical power, and $h_i(t)$ is the impulse response of the ith LED, denoted as (Komine & Nakagawa, 2004):

$$h_i(t) = (m+1)A\cos^m(\theta)\cos(\psi)/(2\pi d^2), \tag{4}$$

where A is the area of the receiver. In our analysis, the Line Of Sight (LOS) and first-order reflection are considered and the spatial resolution corresponding to the LOS and first-order reflection are set as 0.5 m and 1 m, respectively, and reflectivity is set to 0.8. In order to compute SNR, the subcarrier signal during only one time slot is considered. The optical signals simultaneously sent from LEDs at different locations experience individual times of arrival. The power received inside and outside a time slot are considered as signal and Intersymbol Interference (ISI) noise, respectively. They are denoted as $P_{rSignal}$ and P_{rISI}, and are given as (Komine & Nakagawa, 2004):

$$P_{rSignal} = \int_0^{T/4} \left(\sum_{i=1}^{LEDs} h_i(t) \otimes X(t) \right) dt, \tag{5}$$

$$P_{rISI} = \int_{T/4}^{\infty} \left(\sum_{i=1}^{LEDs} h_i(t) \otimes X(t) \right) dt, \tag{6}$$

where for SC-4PPM, X(t) represents the transmitted subcarrier pulse within a time slot. Then the SNR can be expressed as (Komine & Nakagawa, 2004):

$$\text{SNR} = \frac{S}{N} = \left(\gamma^2 P^2_{\text{rSignal}}\right) / \left(\sigma^2_{\text{shot}} + \sigma^2_{\text{thermal}} + \gamma^2 P^2_{\text{rISI}}\right), \tag{7}$$

where σ^2_{shot} and $\sigma^2_{thermal}$ are the power of shot noise and thermal noise, respectively, and their relevant parameters are chosen to be consistent with those in Li et al. (2016).

3 OPTIMIZATION PROCESS AND RESULTS

3.1 The LED ring alignment

For the LED ring alignment, depicted in Figure 1(a), 12 LEDs are placed evenly on a ring, whose center is also the center of the ceiling. The distances between the LEDs to the center are the same, denoted as Radius:

$$\text{Radius} = 2.5 \times f - r, \tag{8}$$

where f-r is the coefficient between [0,1].

Conventionally, the SAHP of every LED is 70°. However, we set their SAHP to be an SAHP-ring in the optimization, and implemented a global optimization toolbox in MATLAB software to determine it, in order to obtain optimal communication uniformity.

As mentioned by Li et al. (2016), we employ the Q-factor of SNR:

$$Q_{SNR} = (E(SNR)) / \left(2\sqrt{var(SNR)}\right), \tag{9}$$

to evaluate the entire SNR performance among all the receiver locations, where E(SNR) and var(SNR) are the mean value and variance of SNR, respectively. Q_{SNR} is able to indicate the SNR performance and its distribution simultaneously. When the SNR mean value is fixed, the higher Q_{SNR} is, the lower the SNR variance is and therefore the more uniform the SNR

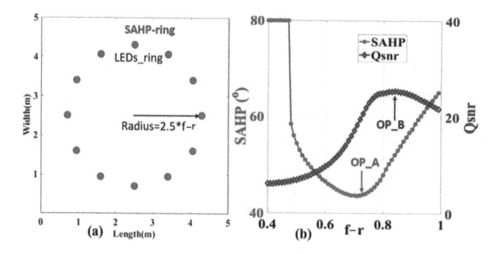

Figure 1. (a) The LED ring alignment; (b) the optimization results of SAHP and Q_{SNR} with respect to f-r.

distribution. When SNR variance is constant, the higher the Q_{SNR} is, and the higher the SNR mean (or the better SNR) is.

Our goal is to achieve optimal SNR performance, so we maximize Q_{SNR} with respect to SAHP-ring and f-r jointly under communication constraints. Hence, the optimization objectivity is denoted as:

$$\begin{bmatrix} \max \ imize : Q_{SNR}(SAHP - ring) \\ s.t \begin{cases} SNR(SAHP - ring) > 13.6 \\ 20° \leq SAHP - ring \leq 80° \end{cases} \end{bmatrix} \qquad (10)$$

In the optimizations, we set f-r from 4.0 to 1.0, with 0.1 interval respectively and implement the optimization.

Figure 1b depicts the optimization results of SAHP-ring and Q_{SNR}. When the f-r is lower than 0.48, the LEDs are placed closely, the optimal SAHP-ring maintained the same 80°, in order to obtain border lighting and hence border communication for the positions near the edge of the room. When the f-r is between 0.48 and 0.71, and LEDs are placed sparsely, so the optimized SAHP-ring decreased in terms of f-r in order to decrease ISI noise reflected from the wall, and it reached the lowest value of 52.2° when f-r is 0.71. Moreover, when the f-r is larger than 0.71, and LEDs are aligned even more sparsely, the optimal SAHP-ring increased in terms of f-r in order to provide border lighting and hence border communication for the position below the center area of the ceiling. The Q_{SNR} is increased in terms of f-r, for f-r lower than 0.84, and then decreased for f-r higher than 0.84. It obtained the maximum of 25.4 when f-r is 0.84, which means the SNR is distributed most evenly for the SAHP-ring optimization methods.

For comparison, we illustrated the lighting and SNR distributions of the Non-Optimized Performance (NOP) situation (f-r = 0.5, SAHP = 70), and that of two turning points, OP-A (f-r = 0.71, SAHP-ring = 43.6°) and OP-B (f-r = 0.84, SAHP-ring = 52.2°), depicted in Figure 1b. For NOP, shown in Figure 2, both the illuminance and SNR maintained their maximum in the center area. However, the performances for the edge positions maintained their minimum. The performance variance between the highest and lowest values are 8.4 dB and 1394 lux for the SNR and illuminance, respectively. The reason for the non-uniformity of NOP is that the LEDs are located closely and the SAHP is wider, which results in inferior uniformity.

The performance of the optimized OP-A is shown in Figure 3. The mean SNR is increased by nearly 4 dB, compared to that of NOP, and meanwhile, the variance of SNR is decreased

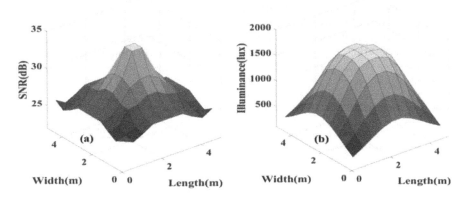

Figure 2. Non-optimized performance: (a) SNR distribution; (b) illuminance distribution.

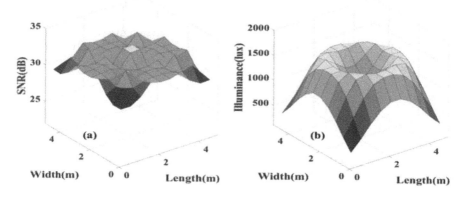

Figure 3. Optimized performance for OP-A situation (f-r = 0.71): (a) SNR distribution; (b) illuminance distribution.

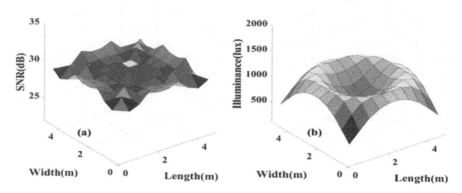

Figure 4. Optimized performance for OP-B situation (f-r = 0.84): (a) SNR distribution; (b) illuminance distribution.

to 3.5 dB. The performance for positions, except for the center area, increased nearly 5 dB. Both the SNR and its uniformity get better, for its sparse LED alignment and narrower SAHP, than for the NOP case. Moreover, the illuminance for the center area decreased more than that of NOP.

The performance of the optimized OP-B is shown in Figure 4. Although the mean SNR decreased to 29.0 dB, compared to the OP-A case by 1.1 dB, the variance of SNR is further decreased by 1.4 dB. It means that the SNR is distributed more uniformly, for its higher f-r and larger SAHP-ring. Furthermore, since Q_{SNR} is the highest among all the other cases,

Table 1. Comparisons of LED ring optimizations and NOP situation.

	f_r	SAHP-ring (deg)	SNR (dB)					Illuminance (lux)		
			Q	min	max	mean	variance	min	max	mean
NOP	0.5	70	6.9	25.0	33.4	27.2	8.4	295	1,689	982
OP-A	0.71	43.6	18.3	29.4	33.0	31.1	3.5	360	1,556	1,167
OP-B	0.84	52.2	25.4	28.2	30.4	29.0	2.1	475	1,117	935

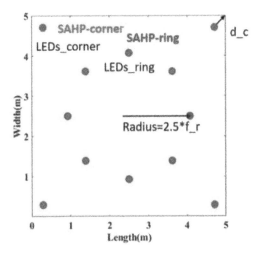

Figure 5.   The LED ring-corner alignment.

the SNR performance produced the most uniform distribution for LED ring alignment. The detailed results of the LED ring alignments are shown in Table 1.

3.2   *The LED ring-corner alignment*

For the LED ring alignment, no matter how the LEDs are located and its SAHP-ring is optimized, the performance of SNR and illuminance for the corners are the worst among all the positions. To increase the corner performance, we locate a single LED in each corner, and set only eight LEDs in the ring, to make sure the total numbers of LEDs are unchanged, as depicted in Figure 5. The distance between the corner and LED is denoted as d-c, and SAHP-corner is the corner LED parameter.

In order to improve the performance of the corner area, the parameters of the corner LEDs, such as d-c and SAHP-corner, are optimized. These parameters, combined with the parameters of the ring LEDs (f-r, SAHP-ring), the four optimization parameters are optimized jointly to obtain optimal SNR uniformity. Hence, the optimization objectivity is denoted as:

$$
\begin{aligned}
&\text{maximize}: Q_{SNR} \\
&(SAHP - ring, SAHP - corner, f - r, d - c) \\
&s.t \begin{cases}
SNR(SAHP - ring) > 13.6 \\
20° \leq SAHP - ring \leq 80° \\
20° \leq SAHP - corner \leq 80°, \\
0 \leq f - r \leq 1 \\
0.1 \leq d - r \leq 1
\end{cases}
\end{aligned}
\tag{11}
$$

The optimized results are 39.8° for SAHP-ring, 48.4° for SAHP-corner, 0.62 for f-r and 0.41 for d-c, as shown in Table 2. Figure 6a and 6b depict the SNR and illuminance distribution employing the optimal ring-corner optimization. Although the Q_{SNR} is slightly lower and the variance of SNR is increased by 0.5 dB more than that of OP-B, the minimum, maximum and the mean values of SNR are higher than that of OP-B, by 1.7 dB, 2.1 dB, and 2.3 dB, respectively. In particular, both the performance of SNR and illuminance near the corner improve. Hence, the overall performance of the optimized LED ring-corner alignment is superior to that of the LED ring alignment.

Table 2. The LED ring-corner optimization results.

	f_r	d_c	SAHP-ring (deg)	SAHP-ring (deg)	SNR (dB)					Illuminance (lux)		
					Q	min	max	mean	variance	min	max	variance
OP-ring-corner	0.62	0.41	39.8	48.4	25.0	29.9	32.5	31.3	2.6	718	1,328	1,050

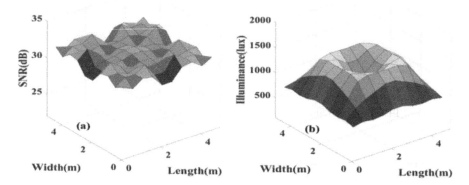

Figure 6. Optimized performance for ring-corner situation (f-r = 0.62, d-c = 0.41, SAHP-ring = 39.8°, SAHP-corner = 48.4°): (a) SNR distribution; (b) illuminance distribution.

4 CONCLUSION

In order to promote the uniformity of the indoor VLC, we optimized the parameters of source alignment and LED SAHPs. For the LED ring alignment, we investigated the ring radius and LED SAHP parameters. Compared to the non-optimized case, the proposed optimization is effective, since the optimized SNR variance can be reduced from 8.4 dB to 2.1 dB, and also the illuminance demonstrates superior uniformity. To further increase the corner performance, we placed four LEDs at the ceiling corner, and optimized the corner LED parameters together with that of the ring LED. The results demonstrate that the overall performance of LED ring-corner alignment is superior to that of LED ring alignment, since the uniformity of SNR and illuminance distribution can both be promoted for positions of corners and edges.

REFERENCES

Borogovac, T., Rahaim, M. & Carruthers, J.B. (2010). Spotlighting for visible light communications and illumination. In *2010 IEEE Globecom Workshops (GC Wkshps)* (pp. 1077–1081).

Boucouvalas, A.C., Chatzimisios, P., Ghassemlooy, Z., Uysal, M. & Yiannopoulos, K. (2015). Standards for indoor optical wireless communications. *IEEE Communications Magazine, 53*(3), 24–31.

Ding, J.P. & Ji, Y.F. (2012). Evolutionary algorithm-based optimisation of the signal-to-noise ratio for indoor visible-light communication utilising white light-emitting diode. *IET Optoelectronics, 6*(6), 307–317.

Higgins, M.D. & Green, R.J. (2015). Optimization of receiving power distribution using genetic algorithm for visible light communication. *Proc. SPIE, 9679*, 967901.

Kim, S. & Jung, S.Y. (2011). Novel FEC coding scheme for dimmable visible light communication based on the modified Reed–Muller codes. *IEEE Photonics Technology Letters, 23*(20), 1514–1516.

Komine, T. & Nakagawa, M. (2004). Fundamental analysis for visible-light communication system using LED lights. *IEEE Transactions on Consumer Electronics*, *50*(1), 100–107.

Komine, T., Tanaka, Y., Haruyama, S. & Nakagawa, M. (2001). Basic study on visible-light communication using light emitting diode illumination. *Proceedings of 8th International Symposium on Microwave and Optical Technology* (pp. 45–48).

Kwon, J.K. (2010). Inverse source coding for dimming in visible light communications using NRZ-OOK on reliable links. *Photonics Technology Letters*, *22*(19), 1455–1457.

Li, F., Wu, K., Zou, W. & Chen, J. (2015). Optimization of LED's SAHPs to simultaneously enhance SNR uniformity and support dimming control for visible light communication. *Optics Communications*, *341*, 218–227.

Li, F., Wu, K., Zou, W. & Chen, J. (2016). Analysis of energy saving ability in dimming VLC systems using LEDs with optimized SAHP. *Optics Communications*, *361*, 86–96.

Noh, J., Lee, S., Kim, J., Ju, M. & Park, Y. (2015). A dimming controllable VPPM-based VLC system and its implementation. *Optics Communications*, *343*, 34–37.

Rajagopal, S., Roberts, R.D. & Lim, S.K. (2012). IEEE 802.15.7 visible light communication: Modulation schemes and dimming support. *IEEE Communications Magazine*, *50*(3), 72–82.

Sung, J.Y., Chow, C.W. & Yeh, C.H. (2014). Dimming-discrete-multi-tone (DMT) for simultaneous color control and high speed visible light communication. *Optics Express*, *22*(7), 7538–7543.

Tanaka, Y., Komine, T., Haruyama, S. & Nakagawa, M. (2003). Indoor visible light transmission system utilizing white LED lights. *IEICE Transactions on Communications*, *86*(8), 2440–2454.

Varma, G.V.S.S.P. (2018). Optimum power allocation for uniform illuminance in indoor visible light communication. *Optics Express*, *26*(7), 8679–8689.

Varma, G.V.S.S.P., Sushma, R., Sharma, V., Kumar, A. & Sharma, G.V.V. (2017). Power allocation for uniform illumination with stochastic LED arrays. *Optics Express*, *25*(8), 8659–8669.

Wang, Z., Yu, C., Zhong, W.D., Chen, J. & Chen, W. (2012). Performance of a novel LED lamp arrangement to reduce SNR fluctuation for multi-user visible light communication systems. *Optics Express*, *20*(4), 4564–4573.

Wu, D., Ghassemlooy, Z., LeMinh, H., Rajbhandari, S. & Kavian, Y.S. (2011). Power distribution and Q-factor analysis of diffuse cellular indoor visible light communication systems. In *2011 16th European Conference on Networks and Optical Communications (NOC)* (pp. 28–31). IEEE.

Optoelectronic devices and integration

Advances in Optoelectronic Technology and Industry Development – Jose & Ferreira (eds)
© 2020 Taylor & Francis Group, London, ISBN 978-0-367-24634-1

Highly efficient and wide-color-gamut organic light-emitting devices based on multi-scale optical design

A. Mikami
Kanazawa Institute of Technology, Kanazawa, Ishikawa, Japan

ABSTRACT: The luminescent properties of organic light-emitting devices have improved with the combination of a nano-sized multi-cathode structure and the external micro-cavity effect. From the detailed optical analysis, it was found that Surface Plasmon (SP) loss in a metal cathode is suppressed to less than 10% due to the combination of long-range and short-range SP. Not less than 90% of optical power in dipole emission can be successfully utilized as propagation light. In addition, it will be shown that the external micro-cavity effect coupled with SP resonance is useful for the improvement of the color purity of the emission. In results, the color gamut in Organic Light-Emitting Devices (OLEDs) approaches the BT.2020 national standard and an external quantum efficiency was improved by a factor of more than 1.5. The effect of the external micro-cavity will be discussed from a viewpoint of multi-scale optical analysis.

1 INTRODUCTION

Organic Light-Emitting Devices (OLEDs) are now widely recognized as a potential application for high-quality flat-panel displays and general lighting. The near 100% internal quantum efficiency of OLEDs has been achieved using phosphorescent or thermally assisted fluorescent materials (Baldo et al., 1999). However, the External Quantum Efficiency (EQE) of conventional devices remains at 20–25% because of poor light extraction efficiency (Tanaka et al., 2007). One of the reasons for this are the quite large losses induced by the Surface Plasmon (SP) polariton, which is the direct interaction between a metal cathode and evanescent wave in near-field of vertical dipole emission (Neyts, 1998). In results, the optical energy of the propagation wave is restricted to a half of the total emission energy. We have found that the light extraction efficiency can be enhanced by using a high refractive index substrate coupled with a micro-lens array and micro-cavity structure (Mikami, 2011). In addition, light extraction efficiency of organic light-emitting devices has improved by using a thin film stacked cathode structure, consisting of semi-transparent metal and an Optical Buffer (OB) layer (Mikami, 2016). In this paper, we propose an external micro-cavity coupled with SP in the cathode, for the improvement of color purity as well as the emission efficiency. The relationship between the external micro-cavity and SP resonance will be discussed from experimental results and theoretical analysis.

2 EXPERIMENTAL METHODS

2.1 Device structure and sample preparation

Figure 1(a) shows a normal device structure, which consists of an Indium-Tin-Oxide (ITO) bottom electrode, a poly(3,4-ethylenedioxythiophene)polystyrene sulfonate (PEDOT:PSS) hole-injection layer, a bis[(1-naphthyl)-N-phenyl]benzidine (NPB) hole-transporting layer, a 4,4'-N,N'-dicarbazole-biphenyl (CBP) Emissive Layer (EML) doped with color-emitting guest, a 2-(4-biphenylyl)-5-(4-tert-butylphenyl-1,3,4-oxadiazole) (Bu-PBD) Electron-Transporting Layer (ETL) and an aluminum cathode. FIrpic, Ir(ppy)₃ and Ir(pic)₂acac are used as color

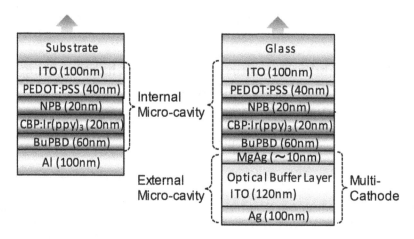

Figure 1. Typical device structures of color-emitting OLEDs used in this experiment: (a) normal structure; (b) multi-layer cathode structure.

dopants for blue, green and red emission. Figure 1b is a proposed device structure with external micro-cavity layers, in which the cathode has three layers consisting of semi-transparent MgAg film, an ITO-based OB layer and high-reflection silver film. In other words, we call it a 'multi-cathode structure'. Each layer was prepared using a physical deposition processes such as vacuum evaporation and rf-sputtering. The polymer layer of PEDOT:PSS was formed using a spin-coating process. The normal device already has a weak micro-cavity between the metal cathode and the high refractive index ITO anode (Bulović et al., 1998). We call it an 'internal micro-cavity'. In addition to that, the Multi-Layer Cathode (MLC) device has the ITO transparent layer sandwiched between two kinds of metals. We call it an 'external micro-cavity'.

2.2 Optical analysis

Optical energy in the OLED is generally divided into a propagation wave and an evanescent wave. The former consists of external, substrate and waveguide modes, and the latter is a direct coupling of near-field with SP mode and lossy surface waves on the metal cathode. These optical modes are shown in Figure 2 as a function of the in-plane wave vector. From the bottom side, the external mode will be treated by classical ray optics. In the case of the substrate mode, we have to take into account a multiple internal reflection in the thick layer. Wave optics of incoherent light will be suitable for this mode. The waveguide mode was calculated from the electro-magnetic optics of coherent light. The SP and lossy surface wave are directly related to the near-field optics. Because a wide range of optics is deeply involved, we performed multi-scale optical analysis by using original simulation software for the optimization of the multi-stacked OLED structure. For the enhancement of light extraction, the most practical way of using a substrate propagation mode is to take the light out of the device by using a high refractive index layer and micro-lens array. However, the sum of external and substrate modes is only less than half of the total optical energy, even if the device is carefully designed, because of large losses induced by waveguide lights and SP (Mikami, 2011). To solve this problem, it will be useful to suppress the SP loss and then convert the waveguide mode into the substrate and external modes.

3 EXPERIMENTAL RESULTS AND DISCUSSION

3.1 Effect of the MLC structure on optical properties

We tried to visualize optical characteristics in three kinds of device structures by using multi-scale optical analysis. Figure 3 shows: [A] optical power density as a function of in-plane wave

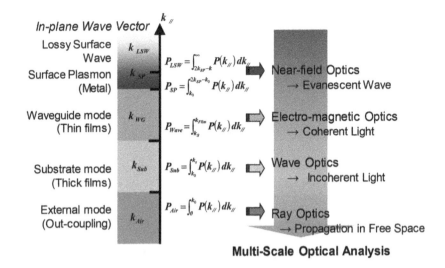

Multi-Scale Optical Analysis

Figure 2. Relationship between optical modes in OLED and optical theories in terms of in-plane wave vector. Because the OLED includes a wide range of wave vectors, a multi-scale model is useful for connecting various optical processes in the OLED.

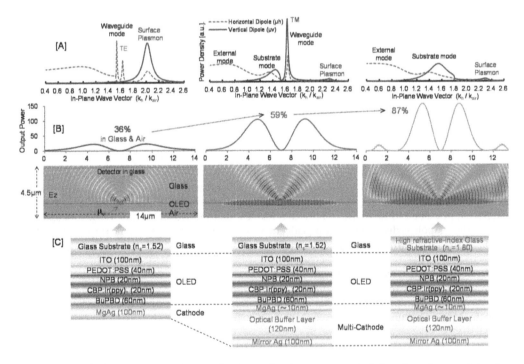

Figure 3. Optical power spectra (power density vs. in-plane wave vector), imaging of power intensity distribution and device structure with different cathode structure and refractive index of glass substrate: (a) normal cathode; (b) multi-layer cathode; (c) multi-layer cathode.

vector (k_h); [B] color images of optical power distribution calculated by Finite-Difference Time-Domain (FDTD) simulation; and [C] device structures with different cathode and substrate; (a) normal cathode and (b), (c) multi-cathode structures. The refractive index (n_s) of

159

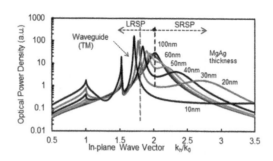

Figure 4. Variation of optical power spectra with the thickness of MgAg layer. The SP band at about 2.0 in k_h is divided into two (LRSP, SRSP) and shift in opposite directions as the MgAg thickness is decreased.

the glass is (a) (b) 1.52 and (c) 1.80, respectively. Since the SP loss mainly originates in the vertical dipole, the power density distribution [B] was calculated only by exciting a vertical dipole moment. In the case of the reference device with normal cathode (a), the emission from the vertical dipole comes out in the direction of slant. Almost all the excitation power is lost and disappears in a short life time, as in the SP in the cathode. In contrast, a strong forward emission can be observed, as well as waveguide mode, in the MLC structure (b). When the high refractive index glass is used as the substrate, the waveguide mode is directly coupled with the substrate mode. The intensity of the pointing vector in the glass increases from 36% to 59% and 87%, suggesting that the sum of the substrate and external modes is increased by 60% by using the MLC structure. As for the power spectrum [A], a strong SP pole appears at 2.02 in the normalized k-space in the reference device. However, it almost disappears and the waveguide Transverse Magnetic (TM) mode becomes dominant in the multi-cathode. A small peak at $k_h = 2.3$ is another SP pole originating in a high-reflection Ag layer because the dipole and mirror is still only 120 nm away. In results, the SP loss decreases to only 12% of the total power. Approximately 90% of the optical energy exists as propagation wave, as a result of the increased waveguide mode. If we use the glass of 1.8 refractive index, TM or Transverse Electric (TE) waveguide lines disappear in the power spectrum. Instead, the substrate mode becomes dominant. This means that the MLC structure makes it possible to convert evanescent wave into propagation wave such as in the waveguide and substrate modes.

Figure 4 shows the variation of power spectra with the thickness of MgAg layer (d_{MgAg}) in the wave vector range of waveguide and SP modes. The refractive index and the thickness of the OB layer were kept constant at 1.8 and 120 nm, respectively. When d_{MgAg} is 100 nm, a strong SP pole appears at 2.02 as well as reference cathode structure. However, when d_{MgAg} is less than 60 nm, the SP pole is divided into two peaks, which is corresponding to Long-Range SP (LRSP) mode in the lower wave vector and Short-Range SP (SRSP) mode in the higher wave vector, respectively. As d_{MgAg} is further decreased to less than 20 nm, LRSP is combined with waveguide mode and changed to propagation light. On the other hand, SRSP disappears because its wave vector exceeds the range of resonance with the SP mode. As a result, almost all of the evanescent wave can be successfully converted to propagation wave. This seems to be the reason why the MLC structure makes it possible to reduce the SP loss.

3.2 *Effect of external micro-cavity on the emission color*

We have experimentally confirmed the effect of external micro-cavity on the device performance of the phosphorescent green OLED doped with Ir(ppy)$_3$. Figure 5 shows a comparison of EQE vs. current density characteristics of the normal and MLC devices with different substrates. High refractive index glass, a half-sphere lens and a micro-lens array sheet were used in order to extract the light out of the substrate in the case of the MLC structure. The reference device with the normal cathode exhibits a maximum EQE of 23% in the low current region. In contrast, MLC devices (b) and (c) show EQEs of 46% and 64% in maximum,

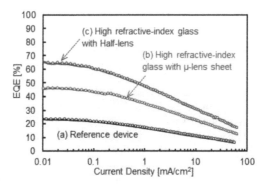

Figure 5. External quantum efficiency vs. current density characteristics in OLEDs with MLC and normal cathode.

respectively. The SP loss is successfully suppressed and converted to waveguide mode by using a multi-cathode structure. The emission efficiency is increased by a factor of two by the combination of high refractive index glass and a micro-lens sheet.

In addition to the green OLED, we tried to make blue and red color emitting devices with Flrpic and Ir(pic)$_2$acac emitting dopants in the MLC device, respectively. Figure 6 shows electroluminescent spectra and color coordinates in a CIE chromaticity diagram in three primary color emitting devices with the normal and MLC structures. The film thickness of the OB layer was adjusted in order to optimize the cavity length between MgAg and Ag films according to the emission wavelength. A sharp emission band with narrower half-width is another advantage in the MLC devices because of the external micro-cavity effect. The triangles of a solid and dashed lines are the color gamut of the MLC and normal structures, respectively. The color region is within the BT.2020 standard, which is indicated by a dotted triangle in the diagram. The color purity of the red, green and blue gamut have been improved greatly by the external micro-cavity effect. The optical energy generated by the MLC structure is coupled with the external micro-cavity. As a result, the cavity structure works effectively, resulting in the improvement of the emission efficiency and color purity.

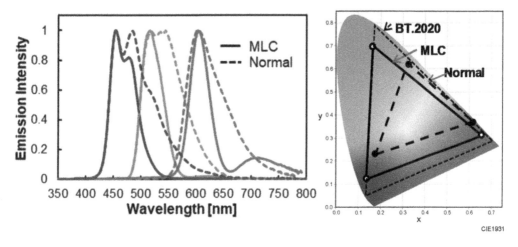

Figure 6. Comparisons of emission spectra and color gamut of three primary colors in OLEDs between the MLC and normal cathodes.

161

4 CONCLUSIONS

The effect of external micro-cavity on the optical properties has been investigated in the color light-emitting OLEDs. The SP loss is successfully suppressed and converted to the waveguide mode by using the MLC structure. It addition, the MLC structure increases the power ratio of the waveguide mode, resulting in the enhancement of micro-cavity effect on the luminescent properties. These phenomena can be explained by the change of wave vector in SP-coupling, induced by the interaction between two SP modes (LRSP and SRSP) on both interfaces of a very thin MgAg layer. The emission efficiency can be increased by the reduction of SP loss. Approximately 60% of optical power in the vertical dipole emission can be converted to glass and air modes. The emission spectra becomes sharp, so the color purity of three primary color emission can be improved by the effect of external micro-cavity and approaches the BT.2020 color standard, which originates in the reduction of SP loss accompanied with increased propagation mode. This optical technique is useful for the improvement of the emission color and efficiency, without sacrificing an electrical property.

REFERENCES

Baldo, M.A., Lamansky, S., Burrow, P.E., Thompson, M.E. & Forrest, S.R. (1999). Very high-efficiency green organic light-emitting devices based on electrophosphorescence. *Applied Physics Letters*, *75*(1), 4–6.

Bulović, V., Khalfin, V.B., Gu, G., Burrows, P.E., Garbuzov, D.Z. & Forrest, S.R. (1998). Weak micro-cavity effects in organic light-emitting devices. *Physical Review B*, *58*(7), 3730–3740.

Mikami, A. (2011). Optical design of 200-lm/W phosphorescent green light emitting devices based on the high refractive index substrate. *Physica Status Solidi (C)*, *8*(9), 2899–2902.

Mikami, A. (2016). Reduction of the optical loss in the multi-cathode structure organic light emitting device using a long range surface plasmon. *Optics and Photonics Journal*, *6*(08), 226–232.

Neyts, K.A. (1998). Simulation of light emission from thin-film microcavities. *Journal of Optical Society of America A*, *15*(4), 962–971.

Tanaka, D., Sasabe, H., Li, Y.J., Su, S.J., Takeda, T. & Kido, J. (2007). Ultra high efficiency green organic light-emitting devices. *Japanese Journal of Applied Physics*, *46(1L)*, L10–L12.

Advances in Optoelectronic Technology and Industry Development – Jose & Ferreira (eds)
© 2020 Taylor & Francis Group, London, ISBN 978-0-367-24634-1

Plastic optical fiber chemosensor for mercury detection in aqueous solution

Junsoo Hong, Hojin Lee & Jaehee Park
Department of Electronic Engineering, Keimyung University, Daegu, Korea

Taeuk Ryu, Min Soo Noh & Sang-Won Park
Department of Environment Science, Keimyung University, Daegu, Korea

ABSTRACT: This paper presents the plastic optical fiber (POF)chemosensor based on an in-line fiber hole and rhodamine derivative for mercury detection in aqueous environments. This sensor is a POF having a rectangular in-fiber hole partially filled with the synthesized rhodamine derivative. The absorbance spectrum of the synthesized rhodamine derivative was changed according to mercury concentration increased. The maximum variations of the absorbance occurred at about 530 nm. Experiments were performed using the POF chemosensor having a 3 mm × 0.65 mm rectangular hole filled with 0.5 mm thickness rhodamine derivative. The transmittance decreased as themercury concentration increased. The experimental results show that the POF chemosensor can be used for detection of mercury ions in aqueous solution.

1 INTRODUCTION

Mercury is a known environmental pollutant routinely released from coal-burning power plant, oceanic and volcanic emission, gold mining, and solid waste incineration. The long atmospheric residence time of mercury vapor and its oxidation to soluble inorganic mercury provide a pathway for contaminating vast amounts of water and soil. Exposure to mercury can be harmful to the brain, heart, kidneys, lungs, and immune system of humans of all ages. The development of detection techniques for real-time and long-term monitoring of mercury contamination in environmental and biological samples has been a high priority (Darbha et al., 2007; Wei et al., 2014).

Standard methods for detecting the presence of mercury in water like cold vapor atomic fluorescence detection and inductively coupled plasma techniques are being employed today (Hanumegowda et al., 2006). However, these methods require bulky and expensive detection equipment, and require a large sample volume and a long process. Optical fiber sensors for mercury detection have been attracted great due to the advantages such as fast measurement time, electromagnetic interference immunity, simplicity, and lower cost (Raj et al., 2016; Kim et al., 2009). Most of optical mercury detection sensors have been the optical chemosensors based on the silica fiber. Although they have high sensitivity, the performance of these sensors are affected seriously by environment noise. Thus, the Plastic Optical Fiber (POF) sensors for detecting mercury in an aqueous environment have received considerable attention (Crosby et al., 2013).

The POF sensors also have some additional merits over glass fiber sensors, such as high elastic strain limits, easy handling, high fracture toughness, high flexibility in bending, high sensitivity to strain, and potential negative thermo-optic coefficients (Park et al., 2015). Thus, a variety of POF sensors have already been successfully applied taking advantage of these properties (Park, 2011; Armin et al., 2011; Ahn. et al., 2019).In this paper, the POF chemosnsor for detection of mercury ion in water based on an in-fiber hole and rhodamine is investigated. The structure of POF chemosensor (Figure 1) is the POF having a rectangular

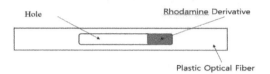

Figure 1. Schematic diagram of POF chemosensor.

type in-fiber hole partially filled with rhodamine derivative. Rhodamine derivative is the chemical material of the POF sensor for detection of mercury ions in water.

2 SENSOR FABRICATION

Rhodamine derivative as a chromophore probe has attracted considerable interest from researchers on account of their excellent photophysical properties (Kim et al., 2008). Rhodamine derivative is colorless, whereas ring-opening of the corresponding spirolactam gives rise to a pink color. When rhodamine binds mercury ions in an aqueous environment, the ring-opening process is induced. Accordingly, the color of rhodamine is changed to pink. In addition, a change in the absorbance spectrum occurs.

In order to synthesize rhodamine derivative, first of all, rhodamine 6G (958 mg, 2 mmol) is dissolved using ethanol (20 mL) and ethylenediamine (0.67 ml, 10 mmol). And this mixture solution is refluxed at the temperature of 70°C for about 20 hours. After refluxing, the precipitate is appeared in the mixture solution. Next, the mixture solution is cooled. Cooled mixture solution is filtered using ethanol and a filter for collecting the precipitate. And then, the collected precipitate is mixed again with acetonitrile solvent (30 ml) to remove impurities and to recrystallize. For making gel-type rhodamine derivatives, the sol-gel method is used. The precipitate is put into tetramethyl orthosilicate (24 ml) solution: trimethoxy methylsilane (6 ml): H_2O (7.5 ml): ethanol (30 ml). Finally, the solution is stirred for 2 hours to make gel-type rhodamine derivative for detecting mercury in an aqueous environment.

The fabrication procedure of the POF chemosensor (Shin et al., 2013) is as follows; first, the rectangular air hole on the POF is fabricated using a microdrilling machine (SMC HD-280) and microdrill bits (NEO Technical System). Next, rhodamine derivative is synthesized according to the above fabrication procedure. Finally, rhodamine derivative is inserted into the rectangular in-fiber hole. The POF chemosensor for detection of mercury ion in water was produced using the POF with a core diameter of 1.48mm, and core and cladding indices of 1.49 and 1.41, respectively, and rhodamine 6G with 95% purity (Avention). The dimension of the rectangular in-fiber hole (Figure 2) was 3 mm × 0.65 mm and the thickness of rhodamine derivative filled inside the in-fiber hole was about 0.5 mm.

Figure 2. Rectangular hole on the POF.

3 EXPERIMENTS AND RESULTS

The experiment setup (Figure 3) for analyzing the characteristics of the POF chemosensor consists of a 532 nm continuous wave laser, a power meter, and a cylinder equipped with the POF sensor. The light from the laser is coupled into the POF and goes to the end of the POF. When the coupling light arrives at rhodamine derivative inside in-fiber hole, the light is absorbed according to the mercury density in aqueous solution. At the end of the POF, the light is detected and the detecting optical power is read by the optical power meter. The optical power detected at the end of the POF was measured while changing the mercury concentration in aqueous solution.

First, the absorbance (Figure 4) of the synthesized rhodamine derivative was measured using a spectrometer (Shimadzu UV-1800). Figure 4 shows that the absorbance varies in the wavelength range from 500 nm to 560 nm with the mercury concentration and the peak absorbance occurs at around 530 nm. The amplitude of absorbance at about 530 nm increases as the mercury concentration increases. This experiment results demonstrate that rhodamine derivatives can be used for detecting mercury ions in an aqueous solution.

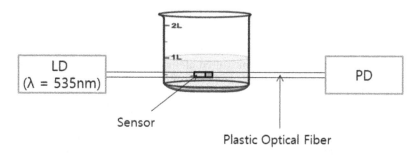

Figure 3.   Experimental setup.

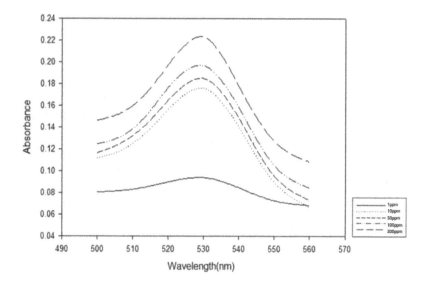

Figure 4.   Absorbance of rhodamine derivative.

3mm+0.5mm

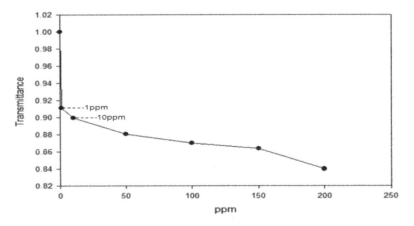

Figure 5.   Transmittance of the POF chemosensor.

After rhodamine derivative was partially filled with the rectangular in-fiber hole in the POF, the transmittances (optical power measured at any ppm/optical power measured at 0 ppm) of the POF chemosensor were measured along with mercury concentration. Figure 5 shows the transmittance of the POF chemosensor measured according to the mercury concentration. The transmittance decreases as the concentration increases. These results explain that the POF chemosensor based on the in-fiber hole and rhodamine derivatives can be used for detection of mercury ions in aqueous solution.

A POF chemosensor for detecting mercury ions in aqueous solution is investigated. This sensor consists of the POF having a 3 mm × 0.65 mm rectangular in-fiber hole partially filled with rhodamine derivative. The absorbance of the rhodamine derivative increased as the mercury concentration increased and the peak absorbance variations occurred at around 530 nm. The transmittance of the POF chemosensor was inversely proportional to the mercury concentration. The experimental results demonstrate that the POF chemosensor can be used to detect mercury ions in aqueous solution.

4   CONCLUSION

This paper presents the POF chemosensor to detect mercury ions in aqueous solution. The POF chemosensor is a POF having a 3 mm × 0.65 mm rectangular hole partially filled with rhodamine derivative. The thickness of rhodamine derivative inside the in-fiber hole was 0.5 mm. The absorbance of rhodamine derivative used as sensing chemical material varied according to the mercury concentration in the wavelength range from 500 to 560 nm and the peak absorbance variations occurred at about 530 nm. The transmittance of POF chemosensor decreased as the mercury concentration increased. The experimental results demonstrate that the POF chemosensor can be used to detect mercury ions in aqueous solution. The dependence analysis of the POF chemosensor performance on solution pH and solution temperature is expected. Furthermore, research to develop a higher-sensitivity POF chemosensor is anticipated.

ACKNOWLEDGMENTS

This work was supported by the National Research Foundation (NRF) of Korea grant funded by the Korean government (MIST) No. 2018R1D1A1B07048066.

REFERENCES

Ahn, D., Park, Y., Shin, J., Lee, J. & Park, J. (2019). Plastic optical fiber respiration sensor based on in-fiber microholes. Microw. Opt. Technol. Lett., 61(1), 120–124.

Armin, A., Stltanolkotabi, M. & Feizollah, P. (2011). On the pH and concentration response of an evanescent field absorption sensor using a coiled-shape plastic optical fiber. Sens. Actuators A, 165(2), 181–184.

Crosby, J.S., Lucas, D. & Koshland, C.P. (2013). Fiber optic based evanescent wave sensor for the detection of elemental mercury utilizing gold nanorods. Sens. Actuators B, 181(5), 938–942.

Darbha, G.K., Ray, A. & Ray, P.C. (2007). Gold nanoparticle-based miniaturized nanomaterial surface energy transfer probe for rapid and ultrasensitive detection of mercury in soil, water, and fish. ACS Nano, 1(3), 208–214.

Hanumegowda, N.M., White, I.M. & Fan, X. (2006). Aqueous mercuric ion detection with microsphere optical ring resonator sensors. Sens. Actuators B, 120(1), 207–212.

Kim, H., Lee, M., Kim, J., Kim, J. & Yoon, J. (2008). A new trend in rhodamine-based chemosensors: Application of spirolactam ring-opening to sensing ions. Chem. Soc. Rev., 37(8), 1465–1472.

Kim, S., Park, J. & Han, W. (2009). Optical fiber AC voltage sensor. Microw. Opt. Technol. Lett., 51(7), 1689–1691.

Park, J. (2011). Plastic optical fiber sensor for measuring driver-gripping force. Opt. Eng., 50(2), 020501.

Park, J., Park, Y. & Shin, J.D. (2015). Plastic optical fiber sensor based on in-fiber microholes for level measurement. Jpn. Appl. Phys., 54(2), 028002.

Raj, D.R., Parsanth, S., Vineeshkumar T.V. & Sudarsanakumar, C. (2016). Surface plasmon resonance based fiber optic sensor for mercury detection using gold nanoparticles PVA hybrid. Opt. Commun., 367(15), 102–107.

Shin, J.D. & Park, J. (2013). Plastic optical fiber refractive index sensor employing an in-line submillimeter hole. IEEE Photon. Technol. Lett., 25(19), 1182–1184.

Wei, Q., Nagi, R., Sadeghi, K., Feng, S., Yan, E., Ki, S., Caire, R., Tseng, D. & Ozcan, A. (2014). Detection and spatial mapping of mercury contamination in water samples using a smart-phone. ACS Nano, 8(2), 1121–1129.

Advances in Optoelectronic Technology and Industry Development – Jose & Ferreira (eds)
© 2020 Taylor & Francis Group, London, ISBN 978-0-367-24634-1

Research of distributed weak fiber Bragg grating sensing system under the action of temperature and strain

Peng Ding, Hongcan Gu, Junbin Huang & Jinsong Tang
Naval University of Engineering, Wuhan, China

ABSTRACT: Effect of temperature on strain measurement in weakly reflective Fiber Bragg Gratings (FBGs)using Time-Division Multiplexing(TDM) is analyzed first.The sensing mechanism of the FBG-TDM is discussed, and strain measurements affected by different kinds of temperature are simulated. Using vibrating liquid column, periodic strain signals are detected in the experiment. The simulated and experimental results show that temperature can distort the strain signal, and the strain signal can be recovered when the temperature influence is reduced. So people should pay more attention to the effect of temperature on strain measurement and try to reduce it.

1   INTRODUCTION

Fiber Bragg Gratings (FBGs) werefirst produced in 1978 (Hill et al., 1978). With rapid progress of manufacturing technology on FBGs (Meltz et al., 1989; Hill et al., 1993; Dong et al., 1993; Martinez et al., 2004), theyare nowwidely used in temperature, strain, displacement, acoustic wave, and magnetic fields (Rao, 1997; Ye et al., 2013; Li, R. et al., 2013; Tao et al., 2017; Tosi, 2018; Kaplan et al., 2019). For large-scale FBG multiplexing, time-division multiplexing (TDM) is used (Morey et al., 1991), and a weakly reflective FBG array (Askins et al., 1994; Bartelt et al., 2007; Yu et al.,2014) is fabricated to further improve its multiplexing capability. Nowadays, systems with weak reflective FBG-TDM based on interference caused by different optical paths areused for earthquake monitoring (Knudsen et al., 2003)and submarine detection (Kirkendall et al.,2007; Nakstad& Kringlebotn,2008; Lin,2013; Lavrov et al., 2017).Vibrational acoustic waves lead to achange of optical path difference, and what should not be neglected is that temperature also has this capability(Wang, Y. et al., 2012; Wang, C. et al., 2018).

It is generally known thatcross-sensitivity of temperature and strain exists in FBGs (Xu et al., 1994). Similar to FBG, effect of temperature on strain measurement ina weakly reflective FBG-TDM system is analyzed first.The laws of temperature and strain measurement are discussed, and strains affected by random, regular temperature are simulated. Two weak reflective FBGs with the same central wavelength in seriesare putin acontainer of vibrating liquid column system.The results of simulation and experiment show that strain measurements are affected by environmenttemperature, and the strain signals are recovered, preferably when the temperature effects are removed by signal processing.

2   SENSING PRINCIPLE

2.1   *System building*

The weak reflective FBG-TDM system is shown in Figure 1. The pulsed laserlaunches periodic pulsed light. Circulators and a 3 × 3 coupler act as light splitters. Symmetric demodulation detects and computes the phase information of the signals (Brawn et al., 1991; Li & Zhang, 2018). The length between G1 and G2 is the same as the distance between Reflector1 and Reflector2 ($L_1 = L_2 = L$). Pulsed light reflected by G2 and Reflector2 and the one that

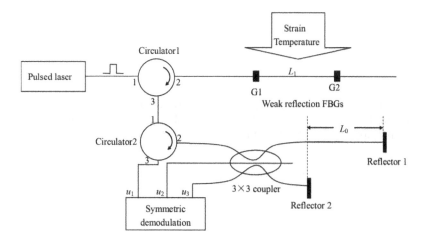

Figure 1.   Block diagram of experimental system.

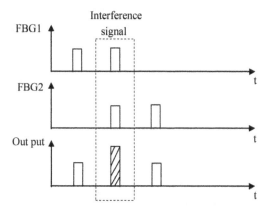

Figure 2.   Time diagram of reflected signals.

pulsed light reflected by G1 and Reflector1 will interfere with each other, andthe interference signal is shown in Figure 2. The phase information of optical path L_1 is shown as follows:

$$\varphi = \frac{4\pi}{\lambda} nL \qquad (1)$$

where φ can be detected by symmetric demodulation; λ is wavelength of light; n is effective refractive index of the fiber. When strain and temperature load to the fiber, n and L will be changed. By measuring φ, the strain and temperature can then be determined.

2.2   Strain and temperature sensing mechanisms

Effective refractive index and length of the fiberdepend on the temperature(T) and on the strain(ε), and their relationships are as follows:

$$\frac{\partial n}{\partial \varepsilon} = -p_e n \qquad (2)$$

$$\frac{\partial L}{\partial \varepsilon} = L \qquad (3)$$

$$\frac{\partial n}{\partial T} = \xi n \tag{4}$$

$$\frac{\partial L}{\partial T} = \alpha_f L \tag{5}$$

From Equations 2 to 5, p_e, ζ, and α_f are the effective elasto-optic coefficient, thermo-optic coefficient, and thermal expansion coefficient of the fiber, respectively. n and L are conducted by Taylor expansion of ε, and the formulas are as follows:

$$n(\varepsilon_0 + \Delta\varepsilon) = n(\varepsilon_0) + \frac{\partial n}{\partial \varepsilon}\Delta\varepsilon + o(\Delta\varepsilon) \tag{6}$$

$$L(\varepsilon_0 + \Delta\varepsilon) = L(\varepsilon_0) + \frac{\partial L}{\partial \varepsilon}\Delta\varepsilon + o(\Delta\varepsilon) \tag{7}$$

In Equations 6 and 7, $o(\Delta\varepsilon)$ arehigh-order terms of $\Delta\varepsilon$, and can be ignored when $\Delta\varepsilon$ is small. From Equations 1, 2, 3, 6 and 7, the change of phase information caused by strain can be calculated as follows:

$$\Delta\varphi = \varphi(\varepsilon_0 + \Delta\varepsilon) - \varphi(\varepsilon_0) \approx \frac{4\pi nL(1 - pe)}{\lambda}\Delta\varepsilon \tag{8}$$

Similarly, n and L are conducted by Taylor expansion of T, and the formulas are as follows:

$$n(T_0 + \Delta T) = n(T_0) + \frac{\partial n}{\partial T}\Delta T + o(\Delta T) \tag{9}$$

$$L(T_0 + \Delta T) = L(T_0) + \frac{\partial L}{\partial T}\Delta T + o(\Delta T) \tag{10}$$

In Equations 9 and10, $o(\Delta T)$ are high-order terms of ΔT and can be ignored when ΔT is small. From Equations 1, 3, 4, 6 and 7, the change of phase information caused by temperature can be calculated as follows:

$$\Delta\varphi = \varphi(T_0 + \Delta T) - \varphi(T_0) \approx \frac{4\pi nL(\alpha_f + \xi)}{\lambda}\Delta T \tag{11}$$

When simultaneous effects of strain and temperature exist, $\Delta\varphi$ can be calculated by Equations 8 and 11 as follows:

$$\Delta\varphi = \frac{4\pi nL(1 - pe)}{\lambda}\Delta\varepsilon + \frac{4\pi nL(\alpha_f + \xi)}{\lambda}\Delta T \tag{12}$$

Here n, L, p_e, α_f, and ζ are assigned as 1.45, 30 m, 0.22, 0.5×10^{-6}, and 6.7×10^{-6}, respectively. If $\Delta\varepsilon = 60 \cos(2\pi t)$, and ΔT are random variations between $-1°C$ and $1°C$, simulation of $\Delta\varepsilon$ detection is shown as in Figure 3a. Figure 3b shows the Signal-to-Noise Ratio (SNR) of the detected signal under different amplitudes of strain. From Figure 3b, the smaller the amplitude of strain, the bigger the influence of short random temperature. In other words, there is a minimum detection limit of strain caused by temperature influence.

If temperature changes regularly, strain signal will be modulated by temperature signal. Figure 4a shows the situation of regular temperature during short time, and Figure 4b shows the situation of regular temperature during long time. From Figure 4a, if the amplitude of strain is small, the detected signal will be broadened and deformed. From Figure 4b, the detected signal will be modulated by a day's temperature change, and the difference between

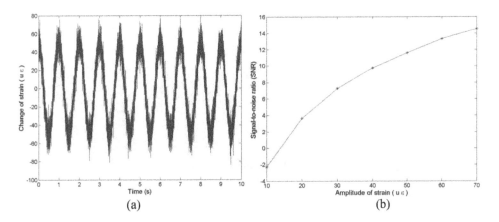

Figure 3. Simulation of $\Delta\varepsilon$ detection affected by short random temperature: (a) $\Delta\varepsilon = 60\cos(2\pi t)$, $\forall\Delta T \in [-1,1]$; (b) signal-to-noise ratio of the detected signal under different amplitude of strain.

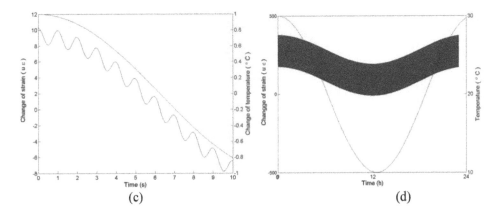

Figure 4. Simulation of $\Delta\varepsilon$ detection affected by regular temperature: (a) During short time; (b) during long time.

minimum and maximum is enlarged, which has a great impact on the dynamic range of system demodulation.

3 EXPERIMENTATION AND ANALYSIS

Figure 5 shows the photograph of the experiment. In the experiment, light source is the narrow linewidth and single frequency laser; the acousto-optical modulator acts as a photoswitch to convert continuous laser into pulsed laser. Liquid column vibration method is used to load strain on two weak reflective FBGs in series, and vibration signals are driven by voltage signals. The light intensity of outputs of 3 × 3 coupler are detected by three detectors respectively, and then the phase information of the light intensity are calculated by symmetric demodulation method.

The weak reflective FBGs are affected by periodic vibrating liquid. The period of vibration is fixed to 5 seconds. The amplitudes of driving voltages are adjusted as 500 mV, 1.5V, and 4V, respectively. Figure 6 shows the detected signals under different driving voltages. From Figure 6, the strain signals are all affected by temperature, and then high-frequency noise and temperature-drift phenomenon are presented. After a de-temperature-drift operation and a normalization operation, the signals are fitted by cosine function, and R-squares of 500 mV, 1.5 Vand 4 Vsituations

171

Figure 5.   Photograph of the experimental setup.

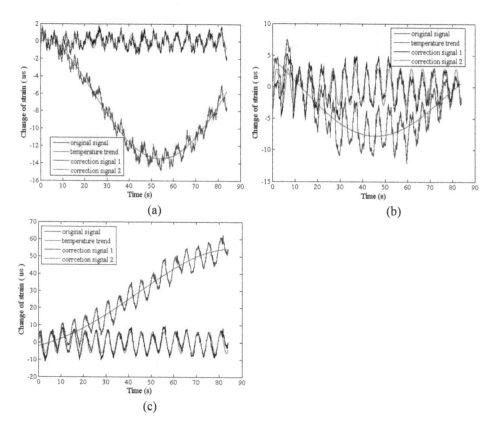

Figure 6.   The detected signals, amplitudes of driving voltages under: (a) 500 mV; (b) 1.5 V; (c) 4 V.

are 0.6338, 0.7333, and 0.8633, respectively; root mean square errors(RMSEs) are 0.2299, 0.2183, and 0.1852. Thus, the greater the vibration amplitude, the easier it is to detect the signal well.

4   CONCLUSIONS

The effect of temperature on vibration signal in weak reflective FBG-TDM system is first ana-lyzed. Its essence is cross-sensitivity of temperature and strain in optical fiber.The distortions of vibration signals by temperature influences are simulated and experimented. The results show that the vibration signals will drift with temperature, which affectsthe dynamic range of system's demodulation, and random temperature duringshorttime will generate high-frequency noise, which limits the minimum detectable strain. Therefore, effective methods must be adopted to reduce the influence of temperature on strain detection, such as using temperature compensation sensors and signal processing.

REFERENCES

Askins, C.G., Putnam, M.A., Williams, G.M. & Friebele, E.J. 1994. Stepped-wavelength optical-fiber Bragg grating arrays fabricated in line on a draw tower. *Optics Letters*19(2):147–149.

Bartelt, H., Schuster, K., Unger, S., Chojetzki, C., Rothhadt, M. & Latka, I. 2007. Single-pulse fiber Bragg gratings and specific coatings for use at elevated temperatures. *Applied Optics* 46(17): 3417–3424.

Brawn, D.A., Cameron, C.B., Keolian, R.M., Gardner, D.L. & Garrett, S.L. 1991. Asymmetric 3 × 3 coupler based demodulator for fiber optic interferometric sensors. *SPIEFiber Optic and Laser Sensors* (1584): 328–335.

Dong, L., Archambault, J.L., Reekie, L., Russell, P. St. J. & Payne, D.N. 1993. Single pulse Bragg gratings written during fibre drawing. *Electronics Letters* 29(17): 1577–1578.

Hill, K.O., Fujii, Y., Johnson, D.C. & Kawasaki, B.S. 1978. Photosensitivity in optical fiber waveguides: application to reflection filter fabrication. *Applied Physics* 32(10): 647–649.

Hill, K.O., Malo, B., Bilodeau, F., Johnson, D.C. & Albert, J. 1993. Bragg gratings fabricated in monomode photosensitive optical fiber photosensitive optical fiber by UV exposure through a phase mask. *Applied Physics Letters*62(1035): 1035–1037.

Kaplan, N., Jasenek, J., Cervenoval, J. & Usakova, M. 2019. Magnetic optical FBG sensors using optical frequency-domain reflectometry. *IEEETransactions on Magnetic* 55(1): 4000704.

Kirkendall, C., Barock, T., Tveten, A. & Dandridge, A. 2007. Fiber otpic towed arrays. *NRL Review* 121–123.

Knudsen, S., Havsgard, G.B., Berg, A., Nakstad, H. & Bostick, T. 2003. Permanently installed high resolution fiber-optic 3C/4D seismic sensor systems for in-well imaging and monitoring applications. *SPIE:The International Society for Optical Engineering* (5278): 51–55.

Lavrov, V.S., Plotnikov, M.Y., Aksarin, S.M., Efimov, M.E., Shulepov, V.A., Kulikov, A.V. & Kireenkov, A.U. 2017. Experimental investigation of the thin fiber-optic hydrophone array based on fiber Bragg gratings. *Optical Fiber Technology* (34): 47–51.

Li, R., Wang, X., Huang, J. & Gu, H. 2013. Spatial-division-multiplexing addressed fiber laser hydrophone array. *Optics Letters* 38(11): 1909–1911.

Li, W. & Zhang, J. 2018. Distributed weak fiber Bragg grating vibration sensing system based on 3 × 3 fiber coupler. *Photonic Sensors* 8(2): 146–156.

Lin, H. 2013. Study on key technologies of the fiber Bragg grating hydrophone array based on path-match interferometry. *China, National University of Defense Technology*.

Martinez, A., Dubov, M., Khrushchev, I. & Bennion. 2004. Direct writing of fibre Bragg gratings by femotosecond laser. *Electronics Letters* 40(19): 1170–1172.

Meltz, G., Morey, W.W. & Glenn, W.H. 1989. Formation of Bragg gratings in optical fibers by a transverse holographic method. *Optics Letters* 14(15): 823–825.

Morey, W.W., Dunphy, J.R. & Meltz, G. 1991. Multiplexing fiber Bragg grating sensors. *SPIEDistributed and Multiplexed Fiber Optic Sensors* (1586): 216–224.

Nakstad, H. & Kringlebotn, J.T. 2008. Realisation of a full-scale fibre optic ocean bottom seismic system. *Proc. of SPIE*(7004):700436.

Rao, Y. 1997. In-fibre Bragg grating sensors. *Measurement Science and Technology* 8(4): 355–375.

Tao, S., Dong, X. & Lai, B. 2017. A sensor for simultaneous measurement of displacement and temperature based on fabry-perot effect of a fiber Bragg grating. *IEEESensors Journal* 17(2): 261–266.

Tosi, D. 2018. Review of chirped fiber Bragg grating(CFBG) fiber-optic sensors and their applications. *Sensors* 18(7): 2147.

Wang, C., Shang, Y., Zhao, W., Liu, X., Wang, C., Yu, H., Yang, M. & Peng, G. 2018. Distributed acoustic sensor using broadband weak FBG array for large temperature tolerance. *IEEE Sensors Journal* 1558.

Wang, Y., Gong, J., Bo, D., Wang, D.Y., Shillig, T.J. & Wang, A. 2012. A larger serial time-division multiplexed fiber Bragg grating sensor network. *Journal of Lightwave Technology* 30(17): 2751–2756.

Xu, M. G., Archambault, J. L., Reekie, L. & Dakin, J. P. 1994. Discrimination between strain and temperature effects using dual-wavelength fibre grating sensors. *Electronics Letters* 30(13): 1085–1087.

Ye, X., Ding, P., Zhou, C., Li, Y., Ni, Y. & Dong, X. 2013. Monitoring of metro-tunnel freezing construction using fiber sensing technology. *Journal of ZhejiangUniversity(Engineering Science)* 47(6): 1072–1080.

Yu, H. & Zheng, Y., Guo, H. & Jiang, D. 2014. Research progress in online preparation techniques of fiber Bragg gratings on optical fiber drawing tower. *Journal of Functional Materials* 45(12): 12001–12005.

Advances in Optoelectronic Technology and Industry Development – Jose & Ferreira (eds)
© 2020 Taylor & Francis Group, London, ISBN 978-0-367-24634-1

Demodulation method for dynamic and static parameters of phase-modulated fiber optical sensor

Shuai Wang, Shun Wang & Ming Zhe Feng
Laboratory of Optical Information Technology, Wuhan Institute of Technology, Wuhan, China

ABSTRACT: This paper proposes a method for demodulating dynamic parameters of fiber-optic sensors, and verifies the feasibility for dynamic parameters of the fiber-optic sensor as well as the demodulation method for static parameters. For the static parameters such as strain or temperature, the optical Vernier structure-based sensors are formed by cascading two single interferometers, and the static parameters can be demodulated by observing the drift of the envelope. For dynamic parameters such as acoustic signal or vibration, by spectral scanning method, the spectrum can be processed to obtain not only the frequency but also the amplitude of the dynamic signal.

1 INTRODUCTION

In recent years, fiber-optic sensors have developed rapidly due to their many merits. In general, optical fiber sensors are mainly classified into phase modulation type sensors (Wu et al., 2017), intensity modulation type sensors (Budiyanto & Suhariningsih, 2017), polarization modulation type sensors (Li, X. et al., 2018), and wavelength modulation type sensors (Liang et al., 2018). Phase modulation sensors have been extensively studied for their high sensitivity. Common phase modulation sensors mainly use four traditional kinds of interference such as Mach-Zehnder interference (Yang et al., 2018), Michelson interference (Zhao et al., 2018), Sagnac interference (Li, X.G. et al., 2018) and Fabry-Perot interference (Li, C. et al., 2018) to transform the change of the external factors into the change of phase. Small changes in external factors can cause changes in the optical path difference (OPD) and the peak wavelength drifts of the interference spectrum. Therefore, the demodulation method for the phase modulation type sensor is usually performed by demodulating the drift amount of the interference peak wavelength. For static parameters such as temperature, refractive index, stress and other parameters, the demodulation method usually uses the optical Vernier effect to amplify the sensitivity, and these parameters are recovered by demodulating the drift of the envelope (Jin et al., 2018; Jiang et al., 2014). For dynamic parameters for example sound signals, the OPD will change synchronously with the frequency of the acoustic signal. In this case, the phase is a variable that changes with time, so the demodulation method under static is no longer applicable. Generally, the methods to demodulate dynamic signal are edge filtering (Gong et al., 2017), phase-generation carrier method (Liu et al., 2017), dual-wavelength orthogonal method (Liao et al., 2017), and so on. However, they have some inevitable drawbacks, i.e. either precise control of certain parameters or complex system is required.

In this paper, a demodulation method of dynamic parameters for fiber-optic sensors is proposed. The feasibility of phase modulation sensors for dynamic parameters and static parameters demodulation is verified by simulation. For the static parameters, the optical Vernier structure is formed by cascading a single interferometer, and the static parameters are demodulated by measuring the envelope drift of the transmission spectrum, and the result shows that the sensitivity is increased by 10 times compared with single interferometer. For dynamic parameters such as acoustic signals, a new spectral scanning method is used to process the spectrum simultaneously demodulate the sound pressure and sound frequency without the need for complex systems or operations.

2 PRINCIPLE

Common optical Vernier effects are typically achieved by cascading two interferometers, one is served as a sensing interferometer and the other as a reference interferometer, with a small difference in the Free Spectral Range (FSR) of the two interferometers. In this paper, a static Vernier parameter is measured by cascading a Mach-Zehnder Interferometer (MZI) and an FP Interferometer (FPI) to form an optical Vernier structure. The FPI consists of a single-mode fiber ceramic ferrule, a polymer film and an aluminum sleeve.

The experimental setup is shown in Figure 1. The incident light is emitted from a broadband source (BBS) through an optical isolator through a MZI consisting of two 1*2 fiber couplers with L1 and L2 arms. The output light is then received by an Optical Spectral Analyzer (OSA) at the FPI connected via a fiber optical circulator. The cavity of the FPI is composed of a fiber end face of a single-mode fiber ceramic ferrule and a polymer film. The cavity length is L3. Transmission spectrum is:

$$I_{out} = I_{in}\left\{\left[A + B\cos\left(\frac{2\pi n_1 \Delta L}{\lambda}\right)\right] * \left[C + D\left(\frac{4\pi n_2 L_3}{\lambda}\right)\right]\right\} \tag{1}$$

where $\Delta L = |L_1 - L_2|$ is the difference between the lengths of the two arms of the MZI; A, B, C, and D are constants. FSR1 and FSR2 are the free spectral ranges of the MZI and the FPI, respectively. Then the FSR of the envelope is as shown in Equation 2. The drifting amount at the peak wavelength of the single FPI output spectrum caused by the acoustic pressure is $\Delta\lambda$. Then, the drift amount at the envelope of the cursor structure output spectrum can be derived in Equation 3:

$$FSR_{envelope} = \frac{FSR_1 * FSR_2}{|FSR_1 - FSR_2|} \tag{2}$$

$$\Delta\lambda_{envelope} = \frac{\Delta\lambda * FSR_1}{|FSR_1 - FSR_2|} \tag{3}$$

When the experimental device is used to measure the acoustic signal, the portion of the MZI is removed. The static spectrum with no sound signal is $S_0(\lambda)$, and when the acoustic signal acts on the sensor element, it is equivalent to loading a ripple spectrum $\Delta S(\lambda, t)$ which determined by wavelength and time on the static spectrum, so the output dynamic spectrum is

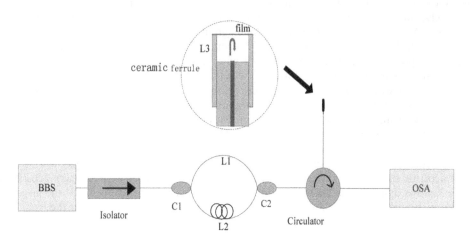

Figure 1. Schematic diagram of the experimental device.

$S(\lambda, t) = S_0(\lambda) + \Delta S(\lambda, t)$. In general, the spectral change (wavelength or intensity) should be linear with the sound pressure (Xu et al., 2014). It is independent of time, so the time variables should be separated from the ripple spectrum. The amplitude of the ripple spectrum can be described as half of the difference between the upper envelope and the lower envelope of the dynamic spectrum. The essence is half of the light intensity difference at that wavelength when the film vibrates to maximize and minimize cavity length. So for a unit of cosine wave with a frequency of ω, the ripple spectrum can be expressed as:

$$\Delta S(\lambda, t) = S(\lambda, t) - S_0(\lambda) = \Delta S(\lambda) \cos(\omega t) = \Delta \lambda_0 * S_0'(\lambda) \cos(\omega t) \tag{4}$$

$S_0'(\lambda)$ is the derivative of the static spectrum, and $\Delta \lambda_0$ is the drift of the peak wavelength of the spectrum with the minimum cavity length when a unit cosine acoustic signal acts on the sensor, and the value has a linear relationship with the acoustic signal amplitude (sound pressure) (Xu et al., 2014).When the amplitude of the acoustic signal changes, the above formula can be converted to:

$$\Delta \lambda * \cos(\omega t) = \frac{S(\lambda, t) - S_0(\lambda)}{S_0(\lambda)} \tag{5}$$

In addition, according to the scanning mechanism of the OSA, the light intensity of each wavelength in the wavelength range can be determined at a certain speed and scanning interval, so the time can be written as the relationship between scanning speed and wavelength $t = \lambda - \lambda_0/V$. In the case of fast Fourier transform directly implemented on the dynamic spectrum (Fu et al., 2018), since the influence of the spatial angular frequency β of the spectrum itself, the frequencies demodulated by the left band and the right band of the interference peak are different, which are $\omega - \beta$ and $\omega + \beta$, respectively. The demodulation frequencies are $\omega - \beta$ and $\omega + \beta$ instead of ω, and the frequency of the acoustic signal can also be obtained by finding the mean value of these two frequencies, increasing reading error and demodulation steps. However, the method mentioned in this paper can eliminate the influence of the interference spectrum itself (eliminating the influence of the spatial angular frequency) by eliminating the envelope of the wavy spectrum with a more effective method, so the demodulated frequency is only ω. Therefore, it is more accurate and convenient.

3 SIMULATION

When the optical Vernier effect is simulated, the FSR_1 and FSR_2 are set to 0.9 nm and 1 nm, respectively, and the $FSR_{envelope}$ is 9 nm from Equation 2. Then the magnification factor $M = \Delta \lambda_{envelope}/\Delta \lambda = 10$. The simulation diagram of the optical Vernier effect is as follows:

In Figure 2, the black line is the spectrum of the MZI, and the blue solid line and the red solid line respectively represent the spectrum when the acoustic pressure acts on the sensor and when it is not applied, and the dotted lines of the corresponding colors indicate the corresponding Vernier spectra. The two orange lines represent the envelope of the two Vernier spectra before and after the acoustic pressure is applied to the sensor. When the peak wavelength of the FPI spectrum caused by the acoustic pressure drifts is 0.2 nm, the drift amount of the envelope of the Vernier structure can be 2 nm. The sensitivity is increased by a factor of 10, which is consistent with the above analysis.

In the simulation of the spectral scanning method, setting w to 30 Hz, select a, b, c three sets of acoustic signals with different amplitudes or frequencies (cos(wt), 2cos(wt), 2cos(2wt)), Figure 3 shows dynamic spectra under three acoustic signals. It can be seen that the larger the sound pressure, the larger the amplitude of the ripple spectrum, and the higher the frequency, the denser the ripple spectrum. It is assumed that the spectral drift caused by the maximum deformation of the thin film due to the sound pressure of a unit cosine sound signal is

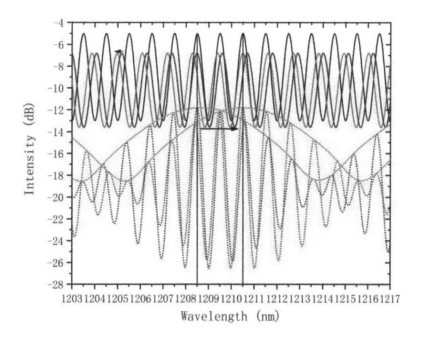

Figure 2.   Schematic diagram of the experimental device.

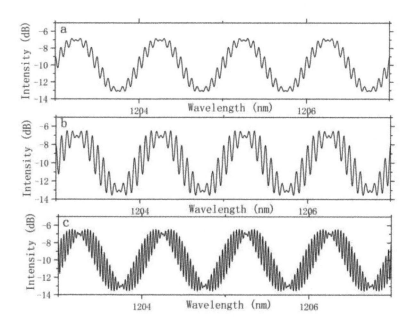

Figure 3.   Dynamic spectrum under different acoustic signals: (a) Interference spectrum when acoustic signal is cos(wt); (b) interference spectrum when acoustic signal is 2cos(wt); (c) interference spectrum when acoustic signal is 2cos(2wt).

$\Delta\lambda_0 = 0.1$ *nm*. Without considering the acoustic frequency response, the spectral drift caused by the acoustic signals of 2cos(wt) and 2cos(2wt) will be 0.2 nm.

The static spectrum of the sound pressure and frequency is obtained by self-fitting the dynamic spectrum. The static spectrum obtained by this method has the same external factor

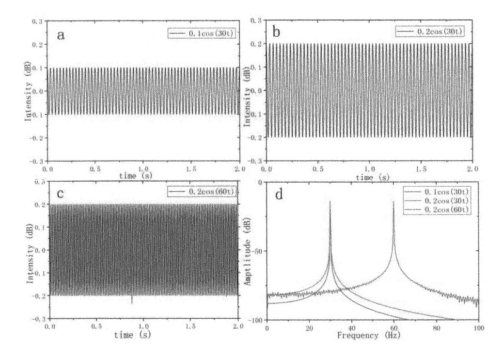

Figure 4. (a) The time-domain spectrum of the demodulated signal under cos(wt); (b) the time-domain spectrum of the demodulated signal under 2cos(wt); (c) the time-domain spectrum of the demodulated signal under 2cos(2wt); (d) frequency domain spectrum of the demodulated signal under the three conditions.

interference as the dynamic spectrum, and can eliminate the influences of external factors when demodulating the acoustic signal by Equation 5. Figures 4a, 4b and 4c are the acoustic signals demodulated according to Equation 5 under these three acoustic signals. It can be seen that the amplitude (represents sound pressure) of the demodulated acoustic signal satisfies linear relationship. Figure 4d is the frequency domain spectrum of the demodulated signal in the three cases. It can be seen that the frequency of the acoustic signal is not affected by the spatial angular frequency, and the contrast is high.

4 CONCLUSION

In this paper, a method for demodulating dynamic signals is proposed, which is applicable but not limited to optical Vernier effect. For acoustic signals, the method can not only demodulate the frequency, but also the sound pressure value of the acoustic signal. In addition, this paper also introduces the demodulation method of the phase-modulated fiber sensor for the static parameters. In general, for the phase modulation fiber sensor, an appropriate demodulation method can be adopted according to the characteristic of the demodulation parameter.

REFERENCES

Budiyanto, M. & Suhariningsih, Y.M. 2017. Cholesterol detection using optical fiber sensor based on intensity modulation. Journal of Physics Conference Series 012008.
Fu, X., Lu, P., Zhang, L., Ni, W.J., Liu, D.M. & Zhang, J.S. 2018. Analysis on Fourier characteristics of wavelength-scanned optical spectrum of low-finesse Fabry-Pérot acoustic sensor. Optics Express 26(17): 22064–22074.

Gong, Z., Chen, K., Zhou, X., Yang, Y., Zhao, Z., Zou, H., & Yu, Q. 2017. High-sensitivity Fabry-Perot interferometric acoustic sensor for low-frequency acoustic pressure detections. Lightwave Technol 35(24): 5276–5279.

Jiang, X., Chen, Y., Yu, F., et al. 2014. High-sensitivity optical biosensor based on cascaded Mach-Zehnder interferometer and ring resonator using Vernier effect. Optics Letters 39(22): 6363–6366.

Jin, Z., Hao, L., Ping, L., et al. 2018. Ultrasensitive temperature sensor with cascaded fiber optic Fabry–Perot interferometers based on Vernier effect. IEEE Photonics Journal 10(5): 1–11.

Li, C., Yu, X., Lan, T., et al. 2018. Insensitivity to humidity in Fabry-Perot sensor with multilayer graphene diaphragm. IEEE Photonics Technology Letters 30(6): 565–568.

Li, X., Ma, R., Xia, Y. 2018. Magnetic field sensor exploiting light polarization modulation of microfiber with magnetic fluid. Journal of Lightwave Technology 36(9): 1620–1625.

Li, X.G., Yong, Z., Xue, Z., et al. 2018. High sensitivity all-fiber Sagnac interferometer temperature sensor using a selective ethanol-filled photonic crystal fiber. Instrumentation Science & Technology 46(3): 253–256.

Liang, H., Jia, P., Liu, J., et al. 2018. Diaphragm-free fiber-optic Fabry-Perot interferometric gas pressure sensor for high temperature application. Sensors 18(4): 1011–1015.

Liao, H., Lu, P.X., Liu, L., Wang S., Ni, W.J., Fu, X., Liu, D., & Zhang, J. 2017. Phase demodulation of short-cavity Fabry–Perot interferometric acoustic sensors with two wavelengths. IEEE Photonics J 9: 1–9.

Liu, B., Lin, J., Liu, H., et al. 2017. Diaphragm based long cavity Fabry–Perot fiber acoustic sensor using phase generated carrier. Opt. Commun. 382: 514–518.

Wu, B., Zhao, C. & Xu, B. 2017. Optical fiber temperature sensor with single Sagnac interference loop based on Vernier effect. Conference on Lasers & Electro-optics Pacific Rim. IEEE 8118631.

Xu, F., Shi, J., Gong, K., Li, H., Hui, R., & Yu, B. 2014. Fiber-optic acoustic pressure sensor based on large-area nanolayer silver diaphragm. Optics Letters 39(10): 2838–2840.

Yang J., Yang, M., Guan, C.Y., et al. 2018. In-fiber Mach-Zehnder interferometer with piecewise interference spectrum based on hole-assisted dual-core fiber for refractive index sensing. Optics Express 26(15): 19091–19099.

Zhao, Y., Ai, Z., Guo, H., et al. 2018. An integrated fiber Michelson interferometer based on twin-core and side-hole fibers for multi-parameter sensing. Journal of Lightwave Technology 36(4): 993–997.

Advances in Optoelectronic Technology and Industry Development – Jose & Ferreira (eds)
© *2020 Taylor & Francis Group, London, ISBN 978-0-367-24634-1*

An improved circulating interferometric integrated optical gyro design method by using graphene-based optical switch

Zhaoyuan Chen, Jun Xu, Yan Gao, Chenjunyi Yang & Zhanrong Zhou
Department of Basic Courses, Rocket Force University of Engineering, Xi'an, China

ABSTRACT: Integrated optical gyroscopes have exerted a tremendous fascination on many researchers for their high sensitivity, miniature size and light weight. In this paper, an improved design method based on a graphene electro-optic switch has been presented to reduce the extra loss of the coupler. Simulation results indicate that the presented modulation method can effectively increase the input power and eliminate extra loss.

1 INTRODUCTION

Optical gyroscopes, which work based on the Sagnac effect (Ezekiel & Balsamo, 1977), are considered to be the next generation inertial navigation sensors for their high sensitivity to angular velocity. The Integrated Optical Gyro (IOG) is one of the most attractive research orientations (Ciminelli et al., 2005; Dell'Olio et al., 2014). The IOG integrates the sensor device into one small and light silicon chip, making the sensor more portable and suitable for complex working environment. However, according to the Sagnac effect, the sensitivity of the optical gyro decreases with the shortening of the optical path length. To resolve this problem and obtain a high-sensitivity optical gyro with small volume, the Circulating Interferometric Integrated Optical Gyro (CIIOG) has been presented (Zhang et al., 2002). This type of optical gyro works like ring laser gyro: it makes light travel several times in the sensing ring, and improves the sensitivity by detecting the optical signal of the nth round. But the coupler in this gyro that inputs the signal in the ring, also outputs a part of optical power in every round, which finally attenuates the output signal power.

In this paper, we presented an optical signal modulating method based on the circulating interferometric optical gyro principle. First, we analyze the power attenuation problem of this type of optical gyro. Then, a graphene-based electro-optic switch has been introduced into this system to replace the traditional coupler to reduce the power attenuation. It is proved feasible and effective by simulation testing.

2 CONFIGURATION AND PRINCIPLES OF CIIOG

The CIIOG consists of a ring resonator coupled with a straight waveguide that inputs and outputs optical signals. Figure1 depicts the configuration of a CIIOG. This structure can make full use of the ring resonator by confining the light to traveling inside the resonator several times. The working principles can be explained as follows: the optical signal, going out from the light source, is split by Y branch and coupled into ring resonator through coupler2 as clockwise (CW) and counterclockwise (CCW) beams. After traveling around the ring, a part of the optical power is coupled out through coupler2, and coupled into a detector through coupler1 to form the first

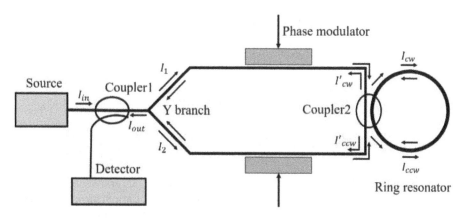

Figure 1. Configuration of CIIOG.

interference signal. The remaining light continues to travel through the ring, and forms the second to nth signal successively.

The detected light intensity can be expressed as $I = I_0(1 + cos(\Delta\varphi))$, where $\Delta\varphi = \varphi_S \pm \pi/2$, $\varphi_S = \Omega 4\pi Lr/\lambda c$ is Sagnac phase shift, and L and r is the length and radius of the ring. The angular velocity Ω can be calculated by $\varphi_S = \Omega 4\pi Lr/\lambda c$.

The nth round optical signal can be expressed as Equations 1 and 2:

$$E_{CCW}^n = \left(t_c\sqrt{1-\alpha^2}\right)t_f^n(t_c\alpha)^{n-1}e^{-i[n\omega(\tau_f-\tau_s-\varphi_0)]}E_{CCW} = R_f e^{-i[n\omega(\tau_f-\tau_s-\varphi_0)]}E_{in} \qquad (1)$$

$$E_{CW}^n = \left(t_c\sqrt{1-\alpha^2}\right)t_f^n(t_c\alpha)^{n-1}e^{-i[n\omega(\tau_f+\tau_s-\varphi_0)]}E_{CCW} = R_f e^{-i[n\omega(\tau_f+\tau_s-\varphi_0)]}E_{in} \qquad (2)$$

where $R_f = (1/\sqrt{2})\left(t_c\sqrt{1-\alpha^2}\right)t_f^n(t_c\alpha)^{n-1}e^{-i[n\omega(\tau_f+\tau_s-\varphi_0)]}$, t_f is the amplitude transmission coefficient, and τ_f and τ_s are the transition time and half the difference between the forward and backward time delays due to rotation, respectively. The two output signals are combined in Y branch and coupled to detector. Neglecting the loss of the Y branch, the optical intensity after interference can be explained as:

$$I_{out}^n = \left|E_{out}^n\right|^2 = \left|E_{CW}^n + E_{CCW}^n\right|^2 \qquad (3)$$

$$I_{out}^n = R_f^2 I_{in}(1 + cos2\omega n\tau_s) = R_f^2 I_{in}(1 + cos\varphi_{ns}) \qquad (4)$$

where φ_{ns} is the Sagnac phase shift in n times the length of the ring. When applied a $\pi/2$ modulation, the output of the system at state of rest is:

$$I_{out}^n = R_f^2 I_{in} = \frac{1}{2}\left(1-\alpha^2\right)^2\alpha^{2(n-1)}t_c^{2n+2}t_f^{2n}I_{in} \qquad (5)$$

Obviously, the total loss of the system can be described by four coefficients: $\left(1-\alpha^2\right)^2$, $\alpha^{2(n-1)}$, t_c^{2n+2}, and t_f^{2n}. The t_f^{2n} represents the amplitude transmission coefficient of the optical path, which depends on the waveguide property. The t_c^{2n+2} is the amplitude transmission coefficient of the coupler. And the $\left(1-\alpha^2\right)^2$ and $\alpha^{2(n-1)}$ represent the input efficiency and output loss at coupler 2. From the discussion above, we can draw the conclusion that the power loss at coupler2 cannot be neglected. To solve this problem, an improved design method that replacing the coupler with a graphene-based electro-optic switch is presented.

3 IMPROVED DESIGN METHOD FOR CIIOG

3.1 *Principles of graphene-based electro-optic switch*

Graphene is a special electrooptical material with high electron mobility (Bolotin et al., 2008), which makes it suitable for high-speed modulating. Its optical property can be characterized by the Kubo formula (Falkovsky, 2008; Chen & Alù, 2011; Gosciniak & Tan, 2013):

$$\sigma_g = i\frac{e^2 k_B T}{\pi \hbar^2 (\omega + i\tau^{-1})}\left[\frac{\mu_c}{k_B T} + 2\ln\left(\exp\left(-\frac{\mu_c}{k_B T}\right) + 1\right)\right] + i\frac{e^2}{4\pi\hbar}\ln\left[\frac{2|\mu_c| - \hbar(\omega + i\tau^{-1})}{2|\mu_c| + \hbar(\omega + i\tau^{-1})}\right] \quad (6)$$

where σ_g is surface conductivity, μ_c is the chemical potential of the graphene films, $\tau = 0.5ps$ represents the momentum relaxation. The surface conductivity of graphene is tuned by chemical potential, which can be changed by bias voltage as follows (Wang et al., 2008):

$$|\mu_c| = \hbar v_F \sqrt{\pi a |V_g - V_{Dirac}|} \quad (7)$$

where v_F is Fermi velocity, and a can be calculated from a basic capacitor model, and $|V_g - V_{Dirac}|$ is the bias voltage because V_{Dirac} approximates to 0. The equivalent permittivity ε_g of a single graphene film can be expressed as $\varepsilon_g = 1 + i\sigma_g/\omega\varepsilon_0 d_g$, where $d_g = 0.34$ nm is the thickness of the graphene layer.

The configuration of our presented graphene-based electro-optic switch is shown in Figure 2. Two 3 dB couplers and two straight waveguides constitute a Mach-Zehnder Interferometer (MZI). One arm of the MZI is manufactured with a graphene-embedded silicon waveguide, that two graphene films are sandwiched by silicon as shown in Figure 3. And the other is

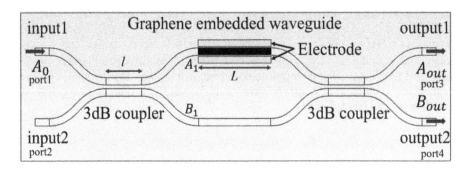

Figure 2.   The configuration of graphene-based electro-optic switch.

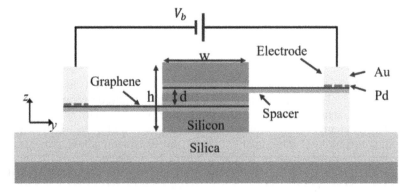

Figure 3.   The layout of the graphene-embedded waveguide.

manufactured only with a silicon waveguide. The graphene films are attached to electrodes to change its optical properties by applying different bias voltages. The embedded graphene films can change the effective index of the waveguide, which will introduce an extra phase that influences the output power ratio among two output ports.

The modulator is manufactured by 500 nm wide and 220 nm high silicon waveguide which is embedded by two graphene films with 10 nm Al_2O_3 spacers. According to Equations 6 and 7, the effective index of the modulator changes with the bias voltage. We calculated the effective index of the graphene-embedded waveguide at different bias voltage by using Lumerical software Mode solution. The results, which are shown in Figure 4, indicate that the real part of the effective index changes rapidly around 5.5 V, where the imaginary part declines dramatically at the same time.

Because the effective index of the silicon waveguide (Neff = 2.00327) doesn't change with bias voltage, the difference of the effective index between the two arms could be tuned by applying bias voltage, which means an extra phase shift ($\Delta\varphi$) is introduced into the upper arm. According to the fundamental theory of waveguides (Okamoto, 2006), the outputs in ports 3 and 4 are:

$$|A_{out}|^2 = |A_0|^2 \cdot sin^2\left(\frac{\Delta\varphi}{2}\right) \tag{8}$$

$$|B_{out}|^2 = |A_0|^2 \cdot cos^2\left(\frac{\Delta\varphi}{2}\right) \tag{9}$$

where $\Delta\varphi = \Delta n \frac{2\pi}{\lambda} L$.

A simulation has been run to research the switch property at different bias voltages. An optical signal is injected from input1, and goes through two 3dB couplers and two arms, and out via output1 and output2. The transmission ratio has been calculated, and its results are shown in Figure 5.

The simulation results indicate that the transmission of optical power from port1 to port3 reaches a maximum of 0.93 when an 8.4 V bias voltage is applied, and the transmission from port1 to port4 reaches a maximum of 0.95 when the applied bias voltage is 19.96 V. In other words, the designed switch is at bar status under 8.4 V bias voltage, and at cross status under 19.96 V bias voltage.

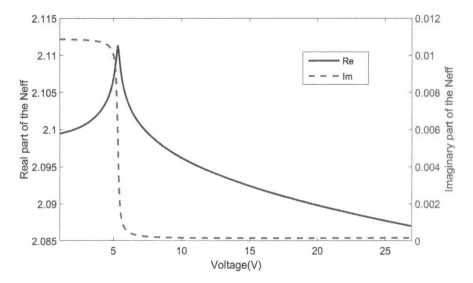

Figure 4.   The effective index of graphene-embedded waveguide at 1550 nm wavelength.

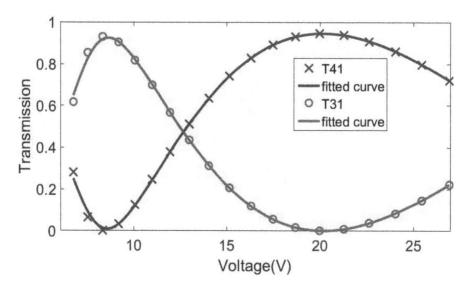

Figure 5.   Transmission vs voltage, where the blue line represents the transmission from port1 to port4, and the red line represents the transmission from port1 to port3.

3.2   *The improved signal modulation method of CIIOG with electro-optic switch*

According to the analysis of CIIOG, most of the power loss is caused by coupler2. Therefore, we proposed to replace the traditional coupler with the designed graphene electro-optic switch, which can change its bar and cross status by applying different bias voltage to alter the coupler status of input and output. The signal modulation method is depicted in Figure 6.

The coupler2 is replaced by a graphene-based electro-optic switch. When the input light propagates to the switch, i.e., moment1 in Figure 6, the switch is altered to cross status until all the optical signal is coupled into the ring at moment2. Then, the switch is altered to bar status to confine the signal to propagating in the ring without extra coupling loss. After traveling n times in the ring, the switch is altered to cross status again from moment3 to moment4, while in this time the first signal goes out and the second goes in, completing the signal input and output.

The optical intensity of the improved method is:

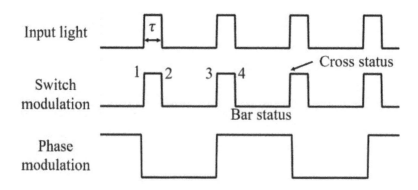

Figure 6.   Modulation scheme of electro-optic switch.

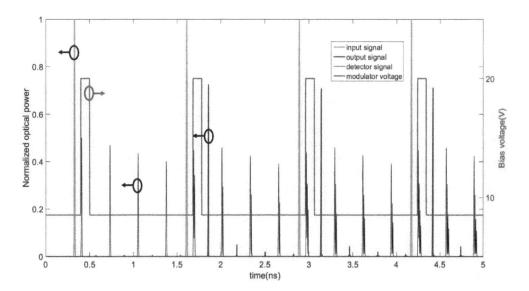

Figure 7. The simulation signals of the CIIOG.

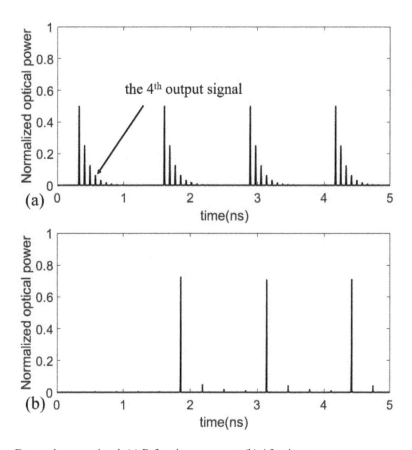

Figure 8. Detected output signal: (a) Before improvement; (b) After improvement.

185

$$I_{out}^{n'} = \frac{1}{2}\left(t_c'\right)^{2n+2} t_f^{2n} I_{in} \tag{10}$$

which is $\left(t_c'/t_c\right)^{2n+2}/\left(1-\alpha^2\right)^2\alpha^{2(n-1)}$ times higher than before.

A simulation has been done to test the designed method. On the basis of the discussion in 3.1, we set the bias voltage at 8.4 V as bar status and at 19.96 V as cross status, which is depicted in Figure 7 by the red line. The green and blue lines in Figure 7 are input and output signals at port1 and port3, which are identified in Figure 2. An optical power detector was placed at A_1 to monitor the signal propagation. Notice that there are four peaks, depicted by black lines, between the two input signals, which means the signal travels four times in the loop before output. The final normalized output signal power is about 0.72.

Figure 8 shows the simulation results of the CIIOG, where (a) is the detected signal before improvement and (b) is the detected signal after improvement. In our simulation, the amount of optical signal circulation is 10. Figure 8a indicates that the optical signal attenuates with each circulation, and the normalized power decays to 0.065 when the signal travels four times in the ring. But Figure 8b shows that, after replacing the coupler with switch, the output normalized power of the tenth circulating signal can still reach up to 0.72, which is 11 times higher than before. This modulation method increases the input optical power, and eliminates the extra coupler loss of order 1 to n-1.

4 CONCLUSION

In this paper, an improved design method of CIIOG has been presented and simulated. We proposed the replacement of the traditional coupler with a graphene-based electro-optic switch to reduce the signal power loss. By changing the bias voltage of the graphene-based electro-optic switch, the coupler status is altered to increase the input signal and eliminate the extra coupler loss. The simulation results prove that the designed method can effectively reduce extra loss caused by the coupler, which has significant implications for the future research of CIIOG.

ACKNOWLEDGMENTS

The authors would like to acknowledge financial support from the National Natural Science Foundation of China (No.61701505) and the Shaanxi Provincial Natural Science Foundation of China (No.2018JQ6080).

REFERENCES

Bolotin, K.I., Sikes, K.J., Jiang, Z., Klima, M., Fudenberg, G., Hone, J., Kim, P. & Stormer, H.L. 2008. Ultrahigh electron mobility in suspended graphene. *Solid State Communications*, 146, 351–355.

Chen, P.-Y. & Alù, A. 2011. Atomically Thin Surface Cloak Using Graphene Monolayers. *ACS Nano*, 5, 5855–5863.

Ciminelli, C., Peluso, F. & Armenise, M.N. 2005. A new integrated optical angular velocity sensor. SPIE.

Dell'olio, F., Tatoli, T., Ciminelli, C. & Armenise, M. 2014. Recent advances in miniaturized optical gyroscopes. *Journal of the European Optical Society - Rapid publications*, 9.

Ezekiel, S. & Balsamo, S.R. 1977. Passive ring resonator laser gyroscope. *Applied Physics Letters*, 30, 478–480.

Falkovsky, L.A. 2008. Optical properties of graphene. *Journal of Physics Conference Series (Online)*, 129, 8.

Gosciniak, J. & Tan, D.T.H. 2013. Theoretical investigation of graphene-based photonic modulators. *Scientific Reports*, 3, 1897.

Okamoto, K. 2006. *Fundamentals of optical waveguides*, Academic Press.

Wang, F., Zhang, Y., Tian, C., Girit, C., Zettl, A., Crommie, M. & Shen, Y.R. 2008. Gate-Variable Optical Transitions in Graphene. *Science*, 320, 206.

Zhang, Y.S., Xiao-Ming, W.U., Tian, W., Tang, Q.A., Tian, Q. & Teng, Y.H.J.J.O.C.I.T. 2002. Circulating interferometric fiber optic gyro and its light source. *Journal of Chinese Inertial Technology*, 10, 45–50.

Advances in Optoelectronic Technology and Industry Development – Jose & Ferreira (eds)
© *2020 Taylor & Francis Group, London, ISBN 978-0-367-24634-1*

A simple frequency-tunable integrated microwave photonic filter based on sideband selective amplification effect

X. Zhang, J. Zheng, T. Pu, Y. Li, J. Li, X. Meng, W. Mou, G. Su, Y. Tan, H. Shi, Y. Chen & T. Dai
College of Communications Engineering, Army Engineering University of PLA, Nanjing, China

S. Ju
The 32298th Troop of Chinese People's Liberation Army, China

ABSTRACT: In this paper, a simple frequency-tunable microwave photonic filter (MPF) is demonstrated and the selective amplification effect has been researched experimentally. The compact MPF is based on an integrated mutual injection DFB laser, which is fabricated by REC technique. The out-of-band rejection ratio is over 30 dB, 3-dB bandwidth is 10 MHz at the frequency of 24.8 GHz. The MPF can be tuned from 16 GHz to 36 GHz by adjusting the bias currents of the laser, which is easy to realize.

1 INTRODUCTION

Processing the microwave signals with high frequency in optical domain has attracted great attention in the past decades compared with more-traditional approaches owing to its superior performance in flexible tunability, low loss, light weight and good immunity to electromagnetic interference. The microwave photonic filter (MPF) is one of the most important approaches in processing microwave signals and provides an effective solution to break the electronic bottleneck of conventional microwave filters, and can find potential applications in radar, optically controlled phased array antennas as well as radio-over-fiber systems (Mora et al., 2006). To obtain wide continuous tunability as well as high resolution, high out-of-band rejection ratio, several approaches have been proposed to realize the MPFs. Capmany et al. (2003) report a novel technical approach that is based on the phase inversion that a RF modulating signal suffers in an electro-optic Mach–Zehnder modulator. A new ultra-wide continuously tunable single-passband MPF based on a stimulated Brillouin scattering technique using a phase modulated optical signal and a dual-sideband suppressed-carrier pump with very high resolution is presented (W. Zhang, 2011). While the scheme has great performance in high resolution with 20-MHz 3-dB bandwidth as well as high out-of-band rejection ratio of 31 dB, the filtering ability at high frequency is not achieved. In order to extend the application of MPFs in high frequency domain, the complex coefficient is implemented by a polarization modulator-based photonic microwave phase shifter, which enables a tunable phase shift from −180° to 180° in 10–40 GHz (Y. M. Zhang, 2013). However, the usage of OBPF and PolM makes the structure more complex and difficult to control because of the polarization variation. Recently, the MPF based on the optoelectronics of semiconductor lasers themselves have been proposed and demonstrated. Xiong and Zhang obtain the MPF based on the principle of wavelength-selective amplification and optical injection (J. Xiong, 2015; Zhang, T., 2016). However, the above systems based on discrete semiconductor lasers, have to employ optical circulator, polarization controller (PC) and/or an external modulator to align the states between the two lasers. Therefore, the entire system suffers from bulky configuration, complexity and instability. Fortunately, the integration of discrete lasers opens a new door to simplify the system configuration.

In this paper, we propose a novel approach to realize a microwave photonic filter which is ultra-sample compared with previously proposed MPFs because the bulky optical injection system is replaced by the integrated mutual injection laser. The tunable MPF is obtained based on optical injection and sideband selective amplification effect which is demonstrated in the experiment. In our previous wok, we found that the MPF is quite an important mechanism which is used as an insert filter in the proposed optoelectronic oscillator (Xin Zhang, 2019). So, we experimentally research the characteristics of the MPF in detail for further applications. The MPF has a wide tunable range from 16 GHz to 36 GHz which is controlled by adjusting the bias currents of the integrated laser. The out-of-band rejection ratio is 30.7 dB when the center frequency of the MPF is 24.8 GHz, and the 3-dB bandwidth is 10 MHz. The corresponding Q factor is 2480.

2 PRINCIPLE

Figure 1 illustrates the experimental diagram of proposed microwave photonic filter. In this system, a compact IMS-DFB laser, which is similar to what has been described in Zheng et al. (2017), is the key device which avoids the traditional bulky optical injection subsystem that is based on discrete optoelectronic devices. The multi-section laser is integrated by three sections including a front DFB section, a phase section and a rear DFB section, which are electrically isolated from each other. While the two DFB sections are optical injected mutually because there is no optical isolator between them. In Figure 1b we can see the three sections can be loaded with different currents respectively for various adjustments and the front laser can be directly modulated. The gratings in both DFB sections are fabricated by REC technique by means of sampled grating pattern which is formed by a conventional holographic exposure combining with a conventional photolithography. Compared with the traditional discrete laser sources that are used for optical injection, it is more challenging to control the wavelengths of the monolithically integrated multi-section laser since all the integrated laser sections share the same heat sink and a single Thermo-Electric Cooler (TEC). Through changing the sampling periods, the wavelengths of the front and rear section lasers and the detuning frequency between the two modes can be controlled accurately. The largest wavelength deviation is less than 0.2 nm which benefits from the high wavelength-controlling accuracy of REC technique, which is sufficient for the process of integrated optical injection.

The operation principle is based on the wavelength-selective amplification of the rear and front section lasers under mutual injection. As is shown in Figure 2a, the cavity modes of the free-running rear and front section lasers are labeled as f_r and f_f. They will red-shift to f_r' and f_f' respectively under mutual injection. There will appear an optical gain spectrum which is centered at the cavity mode f_r' under optical injection. The RF signal f_m is modulated to the front section laser and there are two sidebands on both sides of f_f, as Figure 2b shows. When adjusting I_{DC2}, the weak $+1^{st}$ sideband will fall into the range of the gain spectrum and get amplified in Figure 2c. The rear section cavity mode f_r' will be locked to the amplified side

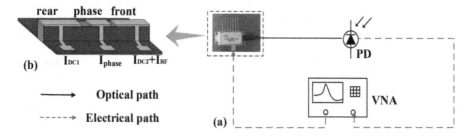

Figure 1. Experimental diagram of the proposed MPF (a) based on the integrated mutual injection DFB laser (b).

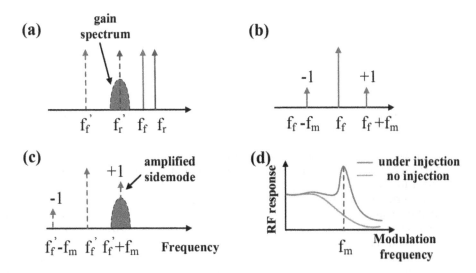

Figure 2. Illustration of the sideband selective amplification effect under mutual injection based on the IMS-DFB laser.

mode if the side mode falls into the locking range. Because it will only happen for those frequencies falling in the gain spectrum, a single passband MPF with its amplitude response shape following the gain spectrum can be generated consequently, as Figure 2d shows.

3 EXPERIMENTAL SETUP AND DISCUSSION

Figure 1a is the structure of the obtained MPF. The light lasing from the IMS-DFB which is modulated with the sweep RF signals from vector network analyzer (VNA Agilent Technologies N523C) is sent to the photodetector (PDFINISAR u2t XPDV2120RA) with the bandwidth of 50 GHz and responsivity of 0.65 A/W. After photoelectric conversion, the obtained RF signal is sent back to the VNA. The displayed S_{21} parameter illustrates the modulation response of the integrated laser. To demonstrate the wavelength selective amplification effect, we set up a structure as shown in Figure 3. The light of the laser which is modulated by the RF signal generated from the RF signal generator (MG3694B 40GHz) is split equally to two parts, 50% is sent to the optical spectrum analyzer (OSAILX-Lightwave, LDC-3724) with the resolution of 0.01 pm and another part is converted to RF signal through the PD which is measured by an Electrical Spectrum Analyzer (ESA). At first, we set the bias current of the rear and front lasers at 85.40 mA and 75.0 mA respectively which makes the laser work at FMW situation. The

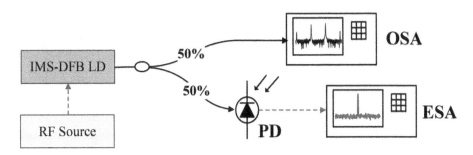

Figure 3. Experimental diagram to demonstrate the selective amplification effect.

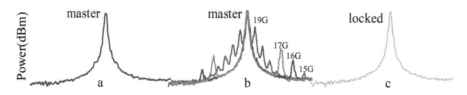

Figure 4. Optical spectrum of the rear section mode with sidebands at different frequencies.

wavelength of the rear laser is shorter than the front one and the beat frequency between them is 19.994 GHz. Then, the RF signals whose power is fixed at 0 dBm with different frequencies are modulated to the front section laser. Figure 4 is the optical spectrum of the rear section mode and its surroundings during the whole experimental process. There is no modulation signal around the rear laser mode before adding the RF signal as is showed in Figure 4a. However, after introducing RF signal whose frequency is larger than 15 GHz, the modulation sideband exists around the master mode. The optical spectrum with RF signal set at 15/16/17/19 GHz respectively is shown in Figure 4b. What's more, as the frequency of the modulation signal increases, which is closer and closer to the master mode, the power of the modulation signal is higher and higher, which means the sideband of the front laser is amplified by the gain spectrum of the master laser mode. In addition, the rear laser mode is locked to the amplified sideband whose frequency is set at 20GHz which is close enough to the rear mode as Figure 4(c) shows. The modulation characteristics get improved under optical injection which is favor to a microwave photonic filter.

Figure 5a is the frequency response of the obtained MPF when the bias currents of rear and front laser are set at 98.97 mA and 85.90 mA, respectively. The out-of-band rejection ratio reaches up to 30.7 dB, which is better than 20 dB in [5]. The injection loss is less than 10 dB which means the desired injection RF signal only undertakes a little bit loss. The insert picture in Figure 5a is the detail of the peak whose center frequency is 24.8 GHz and 3-dB bandwidth is only 10 MHz. Consequently, the corresponding Q-factor is 2480. Through adjusting the bias currents of the integrated laser, the detuning frequency between the two laser modes will vary from 16 GHz to 36 GHz, which means the tuning bandwidth of the obtained MPF is as large as 20 GHz, as Figure 5b shows.

However, the peak response is not flat at different frequencies especially at high frequency. From Figure 5b, we can see the peak decreases about 5 dB compared with that at lower than 25 GHz. We suppose that the sideband amplification effect will reduce when the two modes of the rear and front laser are far away from each other. Specific research will be put into effect in the next work.

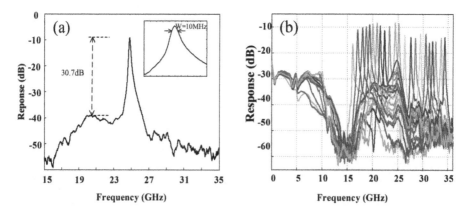

Figure 5. Frequency response of the tunable MPF (b) and details when the center frequency is 24.8 GHz (a), the insert is the 3-dB bandwidth.

4 CONCLUSION

In conclusion, we propose a novel approach to realize a microwave photonic filter which is ultra-sample compared with previously proposed MPFs. The tunable MPF is obtained based on optical injection and sideband selective amplification effect, which is demonstrated in the experiment. The MPF is only composed of an integrated mutual injection DFB laser which is fabricated by REC technology, avoiding employing the external modulator and polarization controllers which are used to align the polarization of the two discrete lasers. The MPF has a wide tunable range from 16 GHz to 36 GHz which is controlled by adjusting the bias currents of the integrated laser. The out-of-band rejection ratio is 30.7 dB when the center frequency of the MPF is 24.8 GHz, and the 3-dB bandwidth is 10 MHz. The corresponding Q factor is 2480. At the same time, the non-flat response at different frequencies will be researched in future work.

REFERENCES

J. Capmany et al., 2003. Microwave photonic filters with negative coefficients based on phase inversion in an electro-optic modulator. *Opt. Lett.*, 28(16): 1415–1417.

J. Mora, B. Ortega, A. Diez, J.L. Cruz, M.V. Andres, J. Capmany, and D. Pastor, 2006. Photonic microwave tunable single-bandpass filter based on a Mach–Zehnder interferometer. *J. Lightw. Technol.*, 24(7): 2550–2509.

J. Xiong et al., 2015. A Novel Approach to Realizing a Widely Tunable Single Passband Microwave Photonic Filter Based on Optical Injection. *IEEE Journal of Selected Topics in Quantum Electronics*, 21(6): 171–176.

Zhang, T., Xiong, J., Zheng, J., et al. 2016. Wideband tunable single bandpass microwave photonic filter based on FWM dynamics of optical-injected DFB laser. *[J]. Electronics Letters*, 52(1): 57–59.

W. Zhang and R.A. Minasian, 2011. Widely tunable single-passband microwave photonic filter based on stimulated Brillouin scattering. *IEEE Photon. Technol. Lett.*, 23(23): 1775–1777.

Xin Zhang, Jilin Zheng et al., 2019. Simple frequency-tunable optoelectronic oscillator using integrated multi-section distributed feedback semiconductor laser. *Opt. Express*, 27(5), 7036–7046.

Y.M. Zhang and S.L. Pan, 2013. Complex coefficient microwave photonic filter using a polarization-modulator-based phase shifter. *IEEE Photon. Technol. Lett.*, 25(2): 187–189.

J. Zheng, G. Zhao et al., 2017. Experimental demonstration of amplified feedback DFB laser with modulation bandwidth enhancement based on the Reconstruction Equivalent Chirp Technique. *IEEE Photonics J.* 9(6), 1–8.

Advances in Optoelectronic Technology and Industry Development – Jose & Ferreira (eds)
© 2020 Taylor & Francis Group, London, ISBN 978-0-367-24634-1

Strained SiGe layer grown on microring-patterned substrate for silicon-based light-emitting devices

Yi Li, Xingzhi Qiu, Chengcong Cui, Jinwen Song, Qingzhong Huang, Cheng Zeng & Jinsong Xia
Wuhan National Laboratory for Optoelectronics, Huazhong University of Science and Technology, Wuhan, Hubei, China

ABSTRACT: A silicon light emitter operating in 1–1.6 μm wavelength range is realized by growth of strained SiGe layer on microring-patterned silicon-on-insulator substrates by molecular beam epitaxy. Strong resonance peaks are observed in the microphotoluminescence spectrum at 5 K and 295 K. The quality factor is of the order of 10^3. The mode indexes and profiles of these whispering-gallery modes are computed through numerical simulation. Significant enhancement of photoluminescence from SiGe layer by microring resonators is attributed to the Purcell effect. Our process provides an enlightening way to fabricate defect-free silicon-based light emitters, and will be further improved in the future.

1 INTRODUCTION

The most important and challenging goal in silicon photonics is to realize silicon-based light sources. In recent years, a silicon-based light source has been realized by bonding (Fang, 2006) or epitaxial III-V materials on a silicon substrate (Chen, 2016). However, due to process complexity and cost considerations, lasers of Group IV (Camacho-Aguilera, 2012) are still very attractive. It is well known that silicon and germanium are indirect semiconductors, which makes it hard to realize lasing. Even so, Ge has been widely investigated for light emitters (Xia, 2007; Wang, 2018; Xu, 2012), in that its luminescence spectrum covers the full telecom-band. One solution to improve light-emission efficiency of Ge is using microcavities such as photonic crystal cavities (Zhang, 2014), microdisks (Xia, 2010), microring resonators (MRRs) (Xia, 2006) and metasurfaces (Yuan, 2017), which make use of the so-called Purcell effect. The light-emission magnitude is enhanced by more than 1000 times when a Ge quantum dot is embedded in a silicon photonic crystal nanocavity (Zeng, 2015) or in metasurfaces (Yuan, 2017). MRRs have been widely used to reduce the threshold of on-chip lasers. In all of these reports, the experimental process is as follows: Ge Quantum Dots (QDs) are grown in MBE or CVD devices, which is followed by pattern transfer by Inductively Coupled Plasma (ICP) or RIE etching. Such an approach would certainly cause the introduction of etch defects in the QDs being etched. Moreover, the Ge QDs in the etched section will be oxidized by oxygen, which introduces further defects, thereby increasing the probability of non-radiative recombination and reducing the light-emission efficiency of the QDs. To avoid the introduction of these defects, the best approach is to make microcavities first, with the Ge layer grown on the patterned substrate afterwards.

In this paper, we report a process to grow a strained SiGe layer on MRR-patterned SOI substrates. The sample morphology is confirmed by Scanning Electron Microscopy (SEM) images. The microphotoluminescence (μPL) of Si MRRs with an embedded SiGe layer at 5 K and 295 K is achieved. A series of sharp resonance peaks are observed in the range from 1 to 1.6 μm. The quality (Q) factor of these resonance peaks ranges from 500 to 1000. Significant enhancement is due to the Purcell effect of the MRRs. The differences between spectra at two temperatures is detailed.

2 DEVICE FABRICATION AND CHARACTERIZATION

We fabricated four groups of MRRs using e-beam lithography (Vistec EBPG 5000 Plus) and ICP etching on a SOI wafer with a 220 nm-thick top silicon film. The internal diameters and widths of the MRRs are listed in Table 1. The substrate is cleaned by the standard RCA method and passivated with H^+ by immersion in dilute HF solution for 80 s before being loaded into a solid-source molecular beam epitaxy chamber (Omicron EVO-50). After degassing at 690°C for 10 min, a 14 nm silicon buffer layer was grown at 400°C. Subsequently, 4.6 monolayers of Ge were deposited and the substrate temperature rose from 450°C to 580°C simultaneously. Finally, a 14 nm Si cap layer was grown at 400°C. Figure 1 shows the SEM images of the 1-μm MRR with a 1 um internal diameter. The width of the MRR has not changed much and is consistent with the designed width because of the short degassing duration and low degassing temperature. No QDs appear on the MRR, although tens of QDs form a 'necklace' at the rim of the residual silicon outside the MRR, which is similar to the growth of Ge QDs on Si microdisks (Wang, 2018). In addition, Ge QDs of high areal density are also observed on the unprocessed region (not shown).

3 RESULTS AND DISCUSSION

The μPL measurement at 5 K is performed in a temperature-controlled liquid-helium cryostat with internal x–y nanopositioners. The sample was excited with a 532 nm diode laser, which was focused to a 2-μm spot on the center of the ring waveguide by a high-NA microscope

Table 1.   Internal diameters and widths of the MRRs.

MRR	Internal diameter (μm)	Width (nm)
1-μm	1.0	550
1.5-μm	1.5	550
2-μm	2.0	600
2.5-μm	2.5	500

Figure 1.   SEM images of the 1-μm MRR; tens of QDs are at the rim of the residual silicon outside the MRR.

objective (NA = 0.85). The μPL signal is collected by the same objective and then dispersed by a monochromator with a 500 mm focus length and recorded by a liquid-nitrogen-cooled InGaAs detector array. Figure 2a shows the μPL spectrum of the 1-μm MRR. These sharp peaks corre spond to the Whispering-Gallery Modes (WGMs) supported by the microring. The Q-factor of the peaks is in the range of 950–1450 and is annotated in Figure 2a. The μPL spectrum of the unpatterned region (UPR) under the same conditions is also presented in Figure 2 and served as a control. The two broad peaks in the range of 1.3–1.6 μm represent direct and indirect bandgap peaks respectively. There is a slump in μPL intensity for wavelength beyond 1580 nm, because the InGaAs detector's detecting wavelength range is 900–1580 nm. Compared with the reference, the μPL intensity of the 1-μm MRR is significantly enhanced at resonant wavelengths. This is mainly attributed to the Purcell effect in MRRs and the increased collection efficiency also has an effect on measured PL intensity. Despite the Purcell effect in MRRs, resonance peaks are not found in the short-wavelength range (1–1.2 μm). This is probably because a large proportion of the photogenerated carriers are consumed by the SiGe layer. On the other hand, consumption of carriers by the SiGe layer in MRRs is faster than that in the UPR (Xia, 2007). Thus, fewer carriers are consumed in the Si region of the MRR than of the UPR, which leads to weaker μPL intensity in the range of 1–1.2 μm.

Figures 2b to 2d show the μPL spectra of the other three MRRs under the same conditions. These spectra are almost the same as that of the first MRR. The Q-factor is much of a muchness. But more resonance peaks are observed on account of a smaller Free Spectrum Range (FSR) at larger diameters of MRR.

We have also obtained μPL spectrum of the 1-μm MRR at 295 K, as shown in Figure 3. The μPL spectrum of the UPR at 295 K is used as a reference. WGMs are still observed at 295 K. It

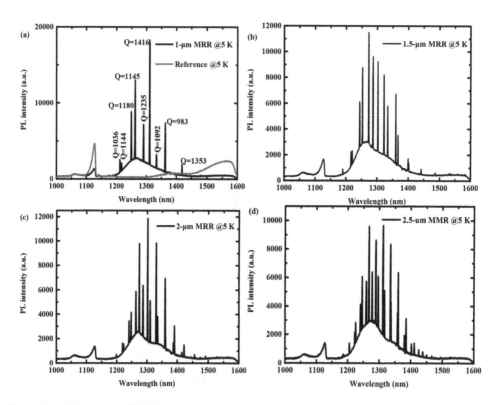

Figure 2. μPL spectra of MRR samples at 5 K: (a) 1-μm MRR [black] and UPR [red], showing Q-factors for the sharpest peaks; (b) 1.5-μm MRR; (c) 2-μm MRR; (d) 2.5-μm MRR.

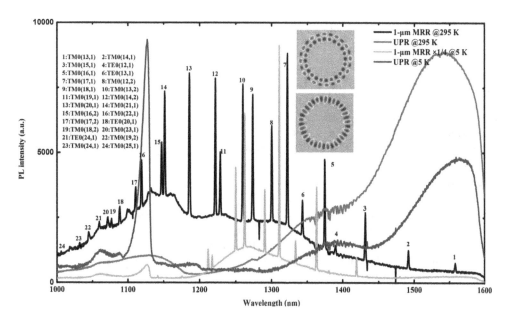

Figure 3. μPL of the 1-μm MRR at 295 K and 5 K; the references are the μPL of the UPR at 295 K and 5 K, together with the mode profiles of TM0(17,1) [bottom] and TM0(12,2) [top] in the inset. The mode indexes of all the WGMs are also labeled and listed.

is noteworthy that resonance peaks appear in the emission range of silicon and one peak is extended to 1558 nm. To explain these differences and investigate the temperature dependence of μPL, the μPL spectra of the 1-μm MRR and UPR at 5 K are also used as references. The former is reduced to a quarter of intensity. What is in accordance with expectation is the red shift of μPL spectrum at 295 K, which is caused by the increase of refractive index (Zeng, 2015) and the thermal expansion of the MRR when temperature increases. However, the Q-factor and the FSRs show little change. A significant reduction in μPL intensity of resonance peaks has been shown when the temperature transfers from 5 K to 295 K, which is caused by the thermal dissociation of excitons at higher temperature (Zeng, 2015). There is less carrier consumption by peaks in the range of 1.2–1.5 μm, so a few weak peaks arise in the range of 1–1.2 μm and 1.5–1.6 μm at 295 K. In the UPR, the μPL in the Si region is broadened, red-shifted and sharply reduced at 295 K. But the μPL intensity from Ge QDs is greatly enhanced. This is because more electrons fill the Γ valley, and more phonons assist indirect transition at 295 K than at 5 K.

The mode indexes TE/TM0(m, n) of all the sharp resonance peaks in the μPL spectrum are labeled and listed in Figure 3. The indexes m and n are azimuthal and radial mode numbers, respectively. Three-Dimensional Finite-Difference Time-Domain (3D-FDTD) simulation is carried out to calculate the mode profile for each WGM of the 1-μm MRR at 295 K with geometry parameters extracted from the SEM image, shown in the insets of Figure 3. As examples, the mode profiles of TM0(17, 1) [bottom] and TM0(12,2) [top] modes are shown in the inset of Figure 3.

Likewise, we measured the μPL spectra of the other three MRRs at 295 K (not shown), which are all similar to that of the 1-μm MRR.

4 SUMMARY

In conclusion, growth of strained SiGe layer on microring-patterned SOI substrate by MBE is demonstrated, which is confirmed by SEM images. μPL from MRRs and UPR is achieved at 5 K and 295 K. Strong resonant peaks with the Q-factor in the range of 950–1450 in μPL spectra from MRRs are observed at both temperature. The Purcell effect is verified when comparing two μPL spectra from MRRs and UPR. What is more, the μPL spectra from MRRs at 295 K is obviously red-shifted and its intensity has fallen a lot. However, the PL intensity from Ge QDs in the UPR is doubled at higher temperature. Because patterning of substrates occurs before Ge deposition, the MRRs are considered to be etch-defect-free and oxygen-defect-free, which is vital to improve the PL intensity of Ge. Besides MRRs, this process can be applied to microdisks and photonic crystal cavities. Considering the small tolerance of metasurface and deformation of pattern during degassing, it seems unfeasible to combine metasurface and site-controlled growth of a Ge layer.

ACKNOWLEDGMENTS

This work was supported by the National Natural Science Foundation of China under Grant Nos. 61335002 and 11574102, and the National High Technology Research and Development Program of China under Grant No. 2015AA016904. I genuinely appreciate the encouragement and comments of He Zhang and Qiang Liu.

REFERENCES

Camacho-Aguilera, R.E., Cai, Y., Patel, N., Bessette, J.T., Romagnoli, M., Kimerling, L.C., & Michel, J. (2012). An electrically pumped germanium laser. *Optics Express*, *20*(10), 11316–11320.

Chen, S., Li, W., Wu, J., Jiang, Q., Tang, M., Shutts, S., ... Smowton, P.M. (2016). Electrically pumped continuous-wave III–V quantum dot lasers on silicon. *Nature Photonics*, *10*(5), 307–311.

Fang, A.W., Park, H., Cohen, O., Jones, R., Paniccia, M.J. & Bowers, J.E. (2006). Electrically pumped hybrid AlGaInAs-silicon evanescent laser. *Optics Express*, *14*(20), 9203–9210.

Wang, S., Zhang, N., Chen, P., Wang, L., Yang, X., Jiang, Z., & Zhong, Z. (2018). Toward precise site-controlling of self-assembled Ge quantum dots on Si microdisks. *Nanotechnology*, *29*(34), 345606.

Xia, J., Takeda, Y., Usami, N., Maruizumi, T., & Shiraki, Y. (2010). Room-temperature electroluminescence from Si microdisks with Ge quantum dots. *Optics Express*, *18*(13), 13945–13950.

Xia, J.S., Ikegami, Y., Shiraki, Y., Usami, N., & Nakata, Y. (2006). Strong resonant luminescence from Ge quantum dots in photonic crystal microcavity at room temperature. *Applied Physics Letters*, *89* (20), 201102.

Xia, J.S., Nemoto, K., Ikegami, Y., Shiraki, Y., & Usami, N. (2007). Silicon-based light emitters fabricated by embedding Ge self-assembled quantum dots in microdisks. *Applied Physics Letters*, *91*(1), 011104.

Xu, X., Narusawa, S., Chiba, T., Tsuboi, T., Xia, J., Usami, N., ... Shiraki, Y. (2012). Silicon-based light-emitting devices based on Ge self-assembled quantum dots embedded in optical cavities. *IEEE Journal of Selected Topics in Quantum Electronics*, *18*(6), 1830–1838.

Yuan, S., Qiu, X., Cui, C., Zhu, L., Wang, Y., Li, Y., ... Xia, J. (2017). Strong photoluminescence enhancement in all-dielectric fano metasurface with high quality factor. *ACS Nano*, 11 (11), 10704–10711.

Zeng, C., Ma, Y., Zhang, Y., Li, D., Huang, Z., Wang, Y., ... Jiang, Z. (2015). Single germanium quantum dot embedded in photonic crystal nanocavity for light emitter on silicon chip. *Optics Express*, *23*(17), 22250–22261.

Zhang, Y., Zeng, C., Li, D., Zhao, X., Gao, G., Yu, J. & Xia, J. (2014). Enhanced light emission from Ge quantum dots in photonic crystal ring resonator. *Optics Express*, *22*(10), 12248–12254.

Advances in Optoelectronic Technology and Industry Development – Jose & Ferreira (eds)
© *2020 Taylor & Francis Group, London, ISBN 978-0-367-24634-1*

A low-cost and compact fiber-optic sensor based on modal interference for humidity sensing

Yun Liu
School of Physics, Dalian University of Technology, Dalian, China

Ping Li, Ning Zhang & Xiaoyong Li
School of Optoelectronic Engineering and Instrumentation Science, Dalian University of Technology, Dalian, China

ABSTRACT: In this work, we proposed a low-cost and simple fiber-optic humidity sensor with a graphene oxide coating. Our sensor was fabricated by splicing single-mode fibers with a short piece of capillary. The sensing region of the sensor was coated by graphene oxide sheets which can absorb water molecule in air and enhance the strength of evanescent fields of the sensor. The humidity sensor was verified by placing it in a humidity chamber at room temperature. With the relative humidity increase, the interference fringe of the sensor shifted to longer wavelength, which indicated the humid environment led to a refractive index change in graphene oxide coating and effected the modal interference of the sensor. This fiber optic humidity sensor has potential to be used for environmental and health monitoring.

1 INTRODUCTION

In recent years, because of small size, remote sensing, electromagnetic immunity and corrosion resistance, fiber-optic humidity sensor has been attracted a great deal of attention and widely used in national defense, scientific research, meteorology and warehousing, etc. (Bernard M. Kulwicki 1991; M. Carolin Mabel & E. Fernandez 2008; Carla Hertleer et al. 2009; Luis Ruiz-Garcia et al. 2009) Fiber-optic humidity sensors of various structures have been reported, including fiber optic humidity sensors based on Fabry-Perot principle (Wei Xu et al. 2013; Weijing Xie et al. 2014; Hao Sun et al. 2015; Dan Suet al. 2013), fiber Bragg grating (FBG) measurement methods (Yao Lin et al. 2015; Pascal Kronenberg et al. 2002), optical fiber Mach-Zehnder (MZI) humidity sensor (Min Shao et al. 2013; Asiah Lokman et al. 2016), long-period fiber grating (LFPG) measurement method (Maria Konstantaki et al. 2006), etc. In general, FO humidity sensors require moisture sensitive materials and transducer mechanisms to achieve humidity measurement. In order to realize the humidity sensing with optical fiber, a necessary step is to modify the fiber structure to allow light leakage. However, the processes are often inseparable from specialized equipment and are destructive to the fiber structure making it more fragile, including fused tapering, misalignment, hydrofluoric acid etching and side polishing and fiber grating (M. Batumalay et al. 2014; Bobo Gu et al. 2011; J. Ascorbe et al. 2016; J. Ascorbe et al. 2017; Getinet Woyessa et al. 2016).

Herein, we demonstrated a graphene oxide coated FO sensor for humidity sensing. The sensing region is consist of guiding fibers and a short piece of capillary. The propagation path of the light covers the entire capillary wall and its evanescent field can permeate through the capillary wall and interact with the GO coating. Because the evanescent field is extremely sensitive to permittivity change of GO coating, this sensor can achieve a high sensitivity relative humidity measurement.

2 FABRICATION AND OPERATION PRINCIPLE

The sensor is manufactured by splicing two single mode fibers to the ends of the capillary (Figure 1a). The capillary has outer diameter and wall thickness of 350 um and 50 um. When the light is injected into the single-mode fiber, the beam entering the capillary rapidly spreads along the curved capillary wall and travels forward in a spiral-like path (Figure 1b). As a result, a variety of higher-order modes can be excited in capillary wall when fundamental mode propagating in the single-mode fiber reaches the capillary. The high-order modes reflect between surfaces of the capillary wall and some of them can enter into the core of lead-out fiber while others are dissipated. Therefore, different modes will meet and interfere with each other when they enter into the lead-out fiber.

Figure 2 shows the humidity measurement setup. The sensor is placed in a humidity chamber and connected with a custom-made fiber-optic sensing interrogator. The interrogator has a resolution of 4pm and a detection range of 1515 nm to 1590 nm. A computer equipped with a data acquisition card was connected to the interrogator for spectral analysis. The detected spectrum was recorded and processed through a program developed based on the Labview software.

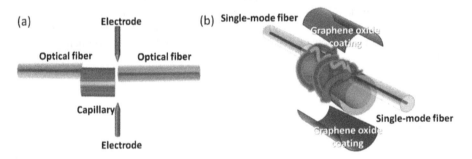

Figure 1. (a) Fabrication of sensor with capillary of 1mm length; (b) schematic of beam propagation in the capillary wall.

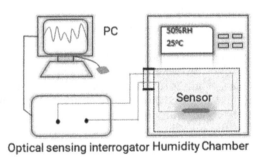

Figure 2. Experimental setup for relative humidity measurement.

3 GRAPHENE OXIDE COATING

As shown in Figure 3, we coated the sensor with graphene oxide film which had many hydrophilic groups. As a moisture sensitive coating, graphene oxide can adsorb water molecules by

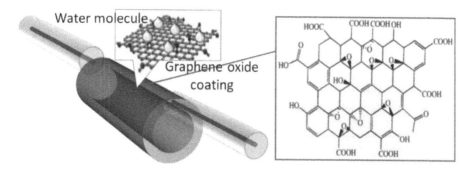

Figure 3. Illustration of the proposed sensor for relative humidity sensing based on GO and evanescent tunneling.

hydrogen bond. Thus, with the relative humidity increases, water molecules will occupy the available active sites of graphene oxide film. Besides, the absorption of water molecules causes a change in the permittivity of the graphene oxide film, which will influences the effective refractive index of the propagating mode in the capillary wall and lead to a wavelength shift of the interference fringe in the transmission of the sensor.

Before coating the graphene oxide film, we ultrasonically cleaned the sensor with deionized water, acetone and isopropanol. Then, the sensor was dried with nitrogen and soaked in piranha solution for 2 hours. A 1 mg/mL graphene oxide aqueous solution was subjected to alkaline treatment with a 0.1 mol/L sodium hydroxide solution to improve the dispersibility of graphene oxide in water. Next, the capillary wall was immersed in graphene oxide aqueous solution to be coated by a dip-coating method. The sensors were examined by scanning electron microscopy as shown in Figure 4. Without graphene oxide coating, the surface of capillary wall was very smooth. As the number of coating cycle increases, more and more wrinkled graphene oxide sheets appeared on the capillary surface.

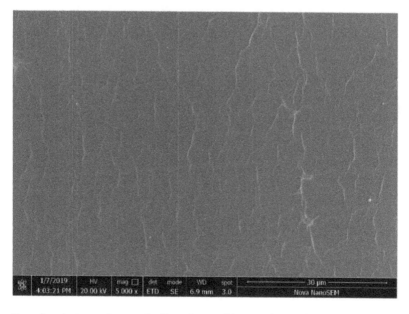

Figure 4. Scanning electron micrograph of graphene oxide coated sensor.

4 RESULTS AND DISCUSSION

We selected a graphene oxide coated sensor with a capillary length of 1mm for the relative humidity measurement. We examined the spectrum of the sensor and found that the interference spectrum degraded with the increase of GO coating thickness. It may be due to the fact that graphene oxide has a higher refractive index compared to air and causes some modes leakage. Figure 5 shows the spectral response of the sensor with 1 cycle and 5 cycles of graphene oxide coating. For the sensor with 1 cycle of graphene oxide coating, the wavelengths of dips on spectrum only slightly fluctuated with the relative humidity changed from 30%RH to 70%RH [Figure 5(a) and 5(c)]. Then, we performed another four cycles of graphene oxide coating and tested the sensor again. As shown in Figure 5(b) and 5(d), dips on spectrum had obviously wavelength shifts to humidity change. Therefore, as the thickness of the graphene oxide coating increased, the sensor gradually showed a wavelength sensitivity to relative humidity. At the same time, because graphene oxide appeared as a high refractive index medium outside the capillary, it caused an increase in light leakage resulting in a degradation of the interference spectrum. From the above experimental results, we can find that the graphene oxide coating has a significant improvement on the humidity sensitivity of the sensor and the influence on the two dips in spectrum is different. This indicates that we can tune the wavelength sensitivity of dips by controlling the amount of graphene oxide coating on the capillary wall.

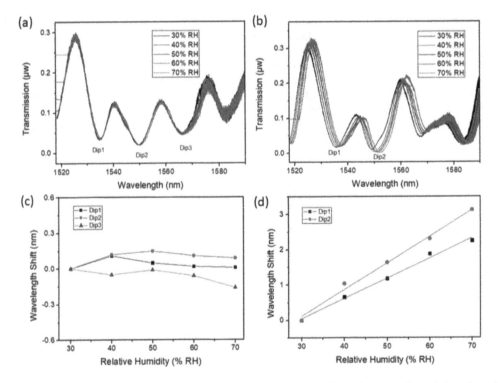

Figure 5. Spectral response of sensors to different relative humidity after 1 cycle and 5 cycles of graphene oxide coating.

5 CONCLUSION

We demonstrated a fiber-optic humidity sensor based on graphene oxide coating which has large surface-to-volume ratio and rich chemical groups. With water molecules absorption by graphene oxide film, the sensitivity of the sensor has been significantly enhanced in humid environment. In the experiment, we applied different cycles of graphene oxide coating on the outside surface of capillary wall. With increase of coating cycle, the sensor became sensitive to the relative humidity change, and its dips had obvious wavelength shift with the increase of relative humidity from 30% RH to 70%RH. As a result, the cycles of GO coating determined the humidity sensitivity of the sensor and the coating process was controllable, which means that the sensor was capable to perform tunable sensitivity.

ACKNOWLEDGEMENT

National Natural Science Foundation of China (Grant No.61705031 and 61727816); China Postdoctoral Science Foundation (Grant No.2017M610175 and 2018T110216).

REFERENCES

Ascorbe, J., Corres, J., Arregui, F., Matias, I. 2017. Recent developments in fiber optics humidity sensors. *Sensors 17*(4): 893.

Ascorbe, J., Corres, J.M., Matias, I.R., Arregui, F.J. 2016. High sensitivity humidity sensor based on cladding-etched optical fiber and lossy mode resonances. *Sensors and Actuators B: Chemical 233*: 7–16.

Batumalay, M., Harith, Z., Rafaie, H.A., Ahmad, F., Khasanah, M., Harun, S.W., Nor, R.M., Ahmad, H. 2014. Tapered plastic optical fiber coated with ZnO nanostructures for the measurement of uric acid concentrations and changes in relative humidity. *Sensors and Actuators A: Physical 210*: 190–196.

Gu, B., Yin, M., Zhang, A.P., Qian, J., He, S. 2011. Optical fiber relative humidity sensor based on FBG incorporated thin-core fiber modal interferometer. *Optics Express 19*(5): 4140–4146.

Hertleer, C., Van Laere, A., Rogier, H., Van Langenhove, L. 2010. Influence of relative humidity on textile antenna performance. *Textile Research Journal 80*(2): 177–183.

Konstantaki, M., Pissadakis, S., Pispas, S., Madamopoulos, N., Vainos, N.A. 2006. Optical fiber long-period grating humidity sensor with poly (ethylene oxide)/cobalt chloride coating. *Applied optics 45*(19): 4567–4571.

Kronenberg, P., Rastogi, P.K., Giaccari, P., Limberger, H.G. 2002. Relative humidity sensor with optical fiber Bragg gratings. *Optics Letters 27*(16): 1385–1387.

Kulwicki, B.M. 1991. Humidity sensors. *Journal of the American Ceramic Society, 74*(4): 697–708.

Lin, Y., Gong, Y., Wu, Y., Wu, H. 2015. Polyimide-coated fiber Bragg grating for relative humidity sensing. *Photonic Sensors 5*(1): 60–66.

Lokman, A., Arof, H., Harun, S. W., Harith, Z., Rafaie, H. A., Nor, R.M. 2015. Optical fiber relative humidity sensor based on inline Mach–Zehnder interferometer with ZnO nanowires coating. *IEEE Sensors Journal 16*(2): 312–316.

Mabel, M.C., & Fernandez, E. 2008. Analysis of wind power generation and prediction using ANN: A case study. *Renewable energy 33*(5): 986–992.

Ruiz-Garcia, L., Lunadei, L., Barreiro, P., Robla, I. 2009. A review of wireless sensor technologies and applications in agriculture and food industry: state of the art and current trends. *Sensors 9*(6): 4728–4750.

Shao, M., Qiao, X., Fu, H., Zhao, N., Liu, Q., Gao, H. 2013. An in-fiber Mach–Zehnder interferometer based on arc-induced tapers for high sensitivity humidity sensing. *IEEE Sensors Journal 13*(5): 2026–2031.

Su, D., Qiao, X., Rong, Q., Sun, H., Zhang, J., Bai, Z., Du, Y., Feng, D., Wang, Y., Hu, M., Feng, Z. 2013.A fiber Fabry–Perot interferometer based on a PVA coating for humidity measurement. *Optics Communications311*: 107–110.

Sun, H., Zhang, X., Yuan, L., Zhou, L., Qiao, X., Hu, M. 2014. An optical fiber Fabry–Perot interferometer sensor for simultaneous measurement of relative humidity and temperature. *IEEE Sensors Journal 15*(5), 2891–2897.

Woyessa, G., Nielsen, K., Stefani, A., Markos, C., Bang, O. 2016. Temperature insensitive hysteresis free highly sensitive polymer optical fiber Bragg grating humidity sensor. *Optics Express 24*(2): 1206–1213.

Xu W., Huang W.B., Huang, X.G., Yu, C.Y.2013. A simple fiber-optic humidity sensor based on extrinsic Fabry–Perot cavity constructed by cellulose acetate butyrate film. *Optical Fiber Technology 19*(6): 583–586.

Xie, W., Yang, M., Cheng, Y., Li, D., Zhang, Y., Zhuang, Z. 2014. Optical fiber relative-humidity sensor with evaporated dielectric coatings on fiber end-face. *Optical Fiber Technology 20*(4):314–319.

Photonics and optoelectronics

Advances in Optoelectronic Technology and Industry Development – Jose & Ferreira (eds)
© 2020 Taylor & Francis Group, London, ISBN 978-0-367-24634-1

Designing and numerical modeling of surface plasmon resonance temperature sensors based on photonic crystal fibers with emphasis on plasmonics and nanophotonics optical quantum metamaterials

Ritu Walia & Kamal Nain Chopra*

Department of Physics, Maharaja Agrasen Institute of Technology, GGSIP University, Rohini, New Delhi, India

*Formerly Laser Science and Technology Centre (LASTEC), DRDO, Metcalfe House, Delhi, India, and Photonics Group, Applied Optics Division, Department of Physics, Indian Institute of Technology, Hauz Khas, New Delhi, India

ABSTRACT: The present paper discusses technically the designing and numerical modeling of surface plasmon resonance temperature sensors based on photonic crystal fibers. Also, the controlling of the temporal properties of electromagnetic radiation by atomically thin graphene monolayer, and metamaterial structures based on it, has been technically discussed. In addition, the organization of the Nan pillars into a 2D periodic array, to form a photonic crystal with quantum dots, has been discussed.

Keywords: surface plasmon resonance temperature sensors, photonic crystal fibers, loss spectra of the core modes, thin graphene monolayer, photonic crystal with quantum dots

1 INTRODUCTION

The use of photonic crystals has recently picked up for use in various optical systems. Chopra (2018a, 2018b) has presented the designing and technical analysis of the use of combination of PhCs based hydrogel with an enzyme hydrogel as biosensors; and the characterization of photonic crystals with emphasis on computation and designing of photonic band structure. In addition, Chopra (2018c) has provided the mathematical designing and short qualitative review of unconventional lasers based on photonic crystals. It is important to note that interest has also been shown recently in the study of the surface plasmon resonance temperature sensor based on photonic crystal fibers. Luan et al. (2014) have proposed a temperature sensor design based on surface plasmon resonances (SPRs) supported by filling the holes of a six-hole photonic crystal fiber (PCF) with a silver nanowire, and filling a liquid mixture (ethanol and chloroform) with a large thermo-optic coefficient into the PCF holes as sensing medium. It has to be noted that the filled silver nanowires can support resonance peaks, and the peak shifts when temperature variations induce changes in the refractive indices of the mixture. It is obvious that the temperature change can be detected by measuring the peak shift. The advantage in this case is that the resonance peak is extremely sensitive to temperature, because the refractive index of the filled mixture is close to that of the PCF material. It has been emphasized that the numerical results indicate that temperature sensitivity as high as 4nm/K can be achieved; and that the most sensitive range of the sensor can be tuned by changing the volume ratios of ethanol and chloroform. Another important point to be noted is that the maximal sensitivity is relatively stable with random filled nanowires, which is clearly very convenient for the sensor fabrication.

It is important to note that the Six-hole PCFs (Wang et al., 2013a), also termed as grapefruit fibers, are suitable for use as sensors, and have been shown in Figure 1. It has been observed that for the optimized sensor operation, each of the holes has to be filled with one silver nanowire of 300 nm diameter, along with a large thermo-optic coefficient liquid. The liquid with silver nanowires has to be filled into the air holes by capillary force and air pressure, and the fabrication should be easy because the air holes of the PCF are quite large.

2   SENSOR DESIGN AND NUMERICAL MODELING

The sensor designing is done by computing and measuring the optical fiber transmission loss $\alpha(dB/cm)$, which is proportional to the imaginary part of the effective index (n_{eff}). The system designer has to make a number of computations to arrive at the optimum design corresponding to the minimum optical fiber transmission loss. This needs a lot of experience on his part, and some times computer software is also required to refine the design. It is important to note that at present, such software is commercially available. To provide an idea about this, the computed loss spectra of the core modes with the refractive index of the liquid at 1.4, as reported in the literature (Luan et al., 2014) have been reproduced below:

It is clear that the resonance peak located at 871 nm is defined by increase in the core mode propagation losses. It may be noted that the losses of a core mode increase dramatically due to the energy transfer into the lossy plasmonic mode. The electric field distributions of the core modes (insets) clearly indicate the energy transferred between the two modes. Also, it is clear that at the non-resonance wavelengths, the core modes (Figure 2, insets (a) and (c)) represent the fundamental mode (HE_{11}), and the energy is found to be mainly confined in the core area. However, at the resonance wavelengths, the core modes and the plasmonic modes become strongly mixed (Figure 2, inset (b)), and hence the energy is transferred into the plasmonic modes. Thus, a peak of the core mode loss spectrum is clearly observed in this wavelength region.

The RI sensitivity of the sensor can be investigated by studying the loss spectra of the core modes in the wavelength range of 700–1400 nm for the different refractive indices of the liquid. The results of the computed loss spectra of the core modes with different refractive indices of the liquid (temperatures); and the resonance wavelength curves of the sensor, when the volume ratio of the ethanol and chloroform is 4:6, as reported in the literature (Luan et al., 2014) have been reproduced in Figure 3.

As is clear, the shift of the resonance peak is from 828 nm for n = 1.39 to 871 nm for n = 1.4, and 928 nm for n = 1.41; which implies that the shift increases with an increase in the value of the refractive index. It can also be understood that the increasing shifts in resonance wavelength

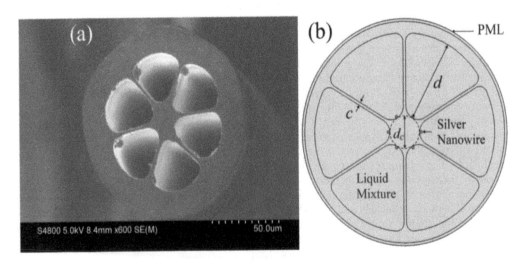

Figure 1.   (a) Cross-section image of a six-hole PCF; (b) Schematic of the proposed sensor fiber. The silver nanowires indicated by megascopic blue dots. Parameters c, d_c and d denote the thickness the core strut, the diameters of the core, and the holes, respectively. Figure courtesy of Wang et al. (2013a).

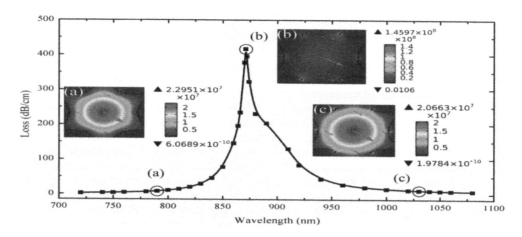

Figure 2.   Calculated loss spectra of the core modes with the refractive index of the liquid at 1.4. Insets show the electric field (E field) distributions of the core modes, and the arrows indicate the polarized direction of electric field. Figure courtesy of Luan et al. (2014).

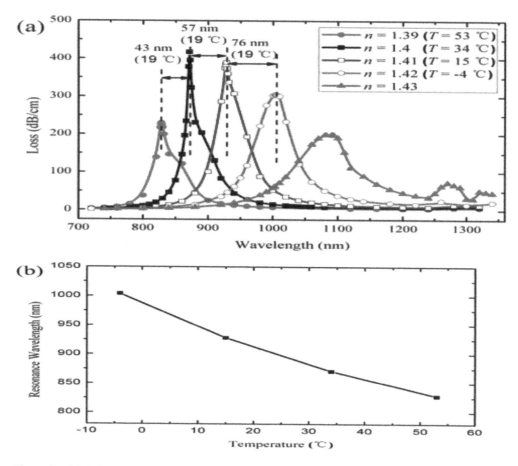

Figure 3.   (a) Calculated loss spectra of the core modes with different refractive indices of the liquid (temperatures); (b) Resonance wavelength curves of the sensor when the volume ratio of the ethanol and chloroform is 4:6. Figure courtesy of Luan et al. (2014).

for the same index change imply the higher sensitivity for the detection range for high n_{liquid} than for that for low n_{liquid}. It has been observed on the basis of the numerical computations that the maximal sensitivity is 7600 nm/Refractive Index Unit (RIU) for n = 1.41–1.42 detection range. Thus, the designer has to optimize these parameters for maximizing the performance of the system. It has to be noted that the phase matching i.e. resonance is achieved by equating the effective refractive indices of the core mode and plasmonic mode at a given wavelength of operation. In fact, the effective refractive index of a core mode is quite close to that of the core material, (n = 1.45 for silica); and the effective refractive index of a plasmonic mode is quite close to that of a bordering liquid. Hence, the high refractive index of the liquid enhances the coupling efficiency between the two modes and the sensitivity. However, interestingly, only one primary peak is observed for the sensor when the refractive index of the liquid is smaller than 1.42; and as the index value increases, secondary peaks appear at long wavelengths. Another important point to be noted is that the secondary peaks may introduce noise and make the detection of high n_{liquid} more difficult; and the increasing index of the liquid results in the low refractive index-contrast of the PCF, which leads to higher losses of the core modes. Hence, the designer has to ensure that the high sensitivity can be achieved only if the refractive index of the filled liquid (the sensing medium) is not more than 1.42. In most of the cases, the designer has to refine the design after feedback from the experimentally achieved data, as the computed data are slightly different from the experimental data.

3 TEMPERATURE SENSITIVITIES OF SENSORS

The sensitivity for temperature sensing applications can be enhanced, by choosing a dielectric material with a high-value thermo-optic coefficient (dn/dT) in the PCF holes, and assuming that a liquid mixture of ethanol and chloroform is filled into the fiber holes as sensing medium (Figure 1b). The reason for the introduction of ethanol is to lower the refractive index of the sensing medium, as the refractive index of chloroform is large (n = 1.44). The refractive index of the liquid is computed by using the following equation (Wang et al., 2013b):

$$n = n_0 + \frac{dn}{dT} \cdot (T - T_0) \tag{1}$$

where n_0 denotes the refractive index of the liquid at the reference temperature T_0. Also, this equation is written by neglecting the material dispersion of the liquid, and assuming n_0 equal to 1.36 for ethanol and 1.44 for chloroform for the spectral regime from 700 nm to 1400 nm at 20 °C. The refractive index of the liquid mixture is computed by using the Lorentz-Lorenz equation (Heller, 1965). For these results, the mixture of ethanol and chloroform with a volume ratio of 4:6 is assumed to be filled into the fiber holes. The loss spectra of core modes and the resonance wavelength curve of the sensor with the mixture at 53 °C, 34 °C, 15 °C, and −4 °C (the mixture still in liquid phase) temperatures have been shown in Figure 3, for which the sensitivity is defined as (Peng et al., 2012):

$$S\lambda[nm/K] = \frac{\Delta\lambda_{peak}}{\Delta T} \tag{2}$$

It has been observed that the maximal sensitivity is 4 nm/K for T = −4–15°C detection range. It is important to note that the most sensitive range of the sensor can be tuned to a desired value by changing the volume ratios of the constituents in the mixture.

It has recently been observed that the atomically thin graphene monolayer and metamaterial structures based on it can be used to tailor and control the spectral, spatial, and temporal properties of electromagnetic radiation. Chen et al. (2012a, 2017b) have reviewed and discussed that the recently discovered two-dimensional (2D) Dirac materials, particularly graphene, can be used as new efficient platforms for excitations of propagating and localized surface plasmon polaritons (SPPs) in the terahertz (THz) and mid-infrared (MIR) regions. They have (i) discussed that the surface plasmon modes supported by the metallic 2D materials exhibit tunable plasmon resonances, which are essential but missing ingredients needed for THz and MIR photonic and optoelectronic devices; and (ii) described how the atomically thin graphene monolayer and metamaterial structures based on it may tailor and control the spectral, spatial, and temporal properties of electromagnetic radiation. It has been stressed that (i) in the same frequency range, the newly unveiled nonlocal, nonlinear, and nonequilibrium electrodynamics in graphene show a number of nonlinear and amplifying electromagnetic responses, whose potential applications are being explored; and (ii) with these 2D material platforms, virtually all plasmonic, optoelectronic, and nonlinear functions found in near-infrared (NIR) and visible devices can be analogously transferred to the long-wavelength regime, even with enhanced tenability, and new functionalities. It has been concluded that the spectral range from THz to MIR is very useful due to the many spectral fingerprints of key chemical, gas, and biological agents, as well as a myriad of remote sensing, imaging, communication, and security applications. Schematic of SPPs in graphene, as used by Chen et al. (2012a, 2017b) have been reproduced in Figure 4.

It may be noted that in the experiment on the manner of launching and detecting the SPPs on graphene, an infrared laser illuminates a metallic tip to locally excite SPPs, and the near-field SPP signal is scattered by the tip into the far-field in order that it may be detected.

5 OPTICAL QUANTUM METAMATERIALS AND METALLIC NANOPARTICLES EXHIBITING COLLECTIVE PLASMONIC MODES

The research work on quantum metamaterials in the optical, or near IR, region of the spectrum has recently picked up, because of the fact that they exhibit plasmatic modes. The term quantum metamaterial is used in various forms to denote: (i) a structure in which quantum degrees of freedom are inserted, (ii) 'quantum dots metamaterial', which is used to stress that, although quantum dots are inserted in a metamaterial, it is not of interest to know about the quantum coherence of the dots, but the gain provided by them is important, for compensating the losses because of the presence of metallic inclusions, and (iii) quantum wells, which are inserted in photonic structures, and are described electromagnetically by a permittivity allowing some control over the behavior of the structure. A common layered metamaterial is one, in which the period consists of two GaAs quantum wells, and results in an effective permittivity tensor allowing obtaining a negative refraction. It is important to note that the effective properties on such structures strongly depend upon the 2D electron density in the quantum well.

Weick and his group (Weick et al., 2013; Weick & Mariani, 2015; Sturges et al., 2015) have discussed that a collection of metallic nanoparticles exhibits collective plasmonic modes. However, it is important to appreciate that a genuine and useful quantum metamaterial, in addition requires the active coherent control of the quantum state of the 'atoms' inserted in the photonic structure, in order to induce a control over the collective properties of the medium. Quach et al. (2009) have introduced the concept of a quantum metamaterial by coupling controllable quantum systems into larger structures, and have stated that the conventional metamaterials represent one of the most important frontiers in optical design, with applications in diverse fields ranging from medicine to aerospace. Initially, the metamaterials had been described as classical structures, which interact only with the classical

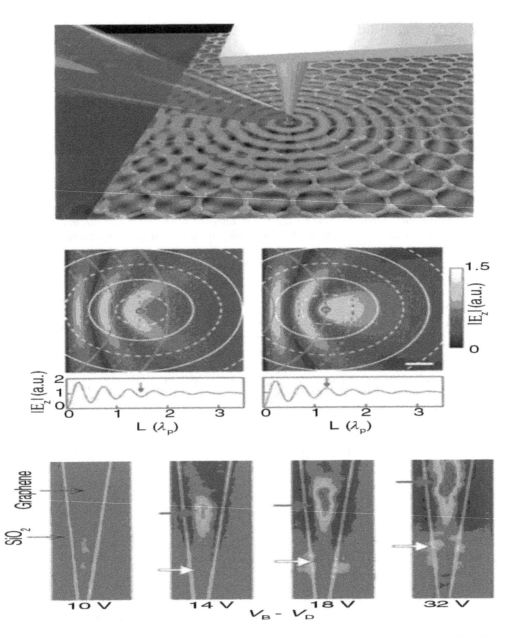

Figure 4. Schematic of SPPs in graphene; (Top) Schematic of the experiment on the manner of launching and detecting the SPPs on graphene based on near-field scanning optical microscopy; (Middle) Snapshots of destructive (left) and constructive (right) interference of SPP waves underneath the tip (blue circles); the profiles of electric field amplitude underneath the tip versus its distance to the left edge, where the arrow indicates the half wavelength of SPPs; (Bottom) Near-field amplitude images of graphene plasmons tuned by an electrostatic backgate. Red and white arrows denote the positions of resonant local SPP modes on a tapered graphene ribbon. Figure modified from figures courtesy of Chen et al. (2012, 2017).

properties of light. Quach et al. (2009) have described a class of dynamic metamaterials, based on the quantum properties of coupled atom-cavity arrays, which are intrinsically lossless, reconfigurable, and operate fundamentally at the quantum level.

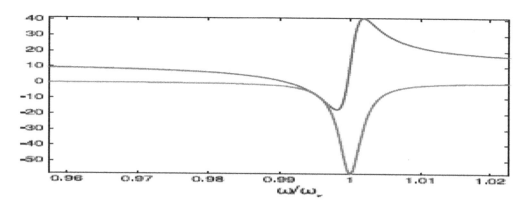

Figure 5. The dielectric function of a core-shell quantum dot. The blue curve is for the real part of the dielectric function, and the red curve is for the imaginary part of the dielectric function. Figure courtesy of Holmström et al. (2010).

It has been discussed that such a system may be implemented by considering a photonic crystal in which quantum oscillators are inserted, e.g. in the form of quantum dots. The quantum dots can be described semi-classically by a dielectric function ε_{QD} (Holmström et al., 2010), as given below:

$$\varepsilon_{QD}(\omega) = \varepsilon_b + \{f_c(E_e) - f_v(E_h)\} \frac{a}{(\omega^2 - \omega^2_0 + 2i\omega\gamma)} \tag{3}$$

where the prefactor $\{f_c(E_e) - f_v(E_h)\}$ represents the difference between the populations of the levels, which can be either positive (in the absorption regime) or negative (in the emission i.e. amplifying regime). The results of the dielectric function of a core-shell quantum dot (Holmström et al., 2010) are illustrated in Figure 5.

It has been suggested (Holmström et al., 2010) that the quantum dots can be grown inside dielectric Nan pillars, and the Nan pillars can be organized into a 2D periodic array, and thus forming a photonic crystal with quantum dots. This is followed by tuning of the bare photonic crystal in such a way that it presents a photonic band gap at the emission frequency of the quantum dots, and thus realizing a pump/probe experiment, where the pump controls the state of the quantum dots (absorption or emission). It may be noted here that when the quantum dots are in the emission regime, a transmission peak appears in the transmission spectrum of the probe. The designer has to ensure that in order to achieve the optimum results, the transmission peak appears in the transmission spectrum of the probe.

6 CONCLUDING REMARKS

The studies on the surface plasmon resonance temperature sensors based on photonic crystal fibers with emphasis on plasmonics and nanophotonics optical quantum metamaterials have recently drawn the attention of various researchers, due to the fact that such materials have applications in research and fabrication of sensors. Efforts are now concentrated on improving the performance and efficiency of the systems based on these materials. Hence, it may be concluded that the field of surface plasmon resonance temperature sensors is evolving fast, and is on a firm footing.

ACKNOWLEDGMENTS

The authors are grateful to Dr. Nand Kishore Garg, Chairman, Maharaja Agrasen Institute of Technology, GGSIP University, Delhi for providing the facilities for carrying out this research work, and also for his moral support. The authors are thankful to Dr. M. L. Goyal, Vice Chairman for encouragement. Thanks are also due to Dr Neelam Sharma, Director, and Dr. V. K. Jain, Deputy Director for their support during the course of the work.

REFERENCES

Chen Pai-Yen, Argyropoulos Christos, Farhat Mohamed, & Gomez-Diaz J. Sebastian, Flatland plasmonics and nanophotonics based on graphene and beyond, Nanophotonics, 6 (2017) 1239–1262.

Chen Jianing, Badioli Michela, Alonso-González Pablo, Thongrattanasiri Sukosin, Huth Florian, Osmond Johann, Spasenović Marko, Centeno Alba, Pesquera Amaia, Godignon Philippe, Elorza Amaia Zurutuza, Camara Nicolas, de Abajo F. Javier García, Hillenbrand Rainer, & Koppens Frank H.L., Optical nano-imaging of gate-tunable graphene plasmons. 487 (2012)77–81.

Chopra Kamal Nain, Designing and Technical Analysis of the Use of Combination of PhCs based Hydrogel with an Enzyme Hydrogel as Biosensors, Proceedings of the 11th International Symposium on Photonics and Optoelectronics SOPO (2018) 89-95, Kunming, China, Taylor and Francis.

Chopra Kamal Nain, A detailed overview and technical analysis of the Photonic Crystals and their characterization with emphasis on computation and designing of Photonic Band structure Atti della "Fondazione Giorgio Ronchi" Anno, ITALY, 73 (2018) 177-215.

Chopra Kamal Nain, Mathematical Designing and Short Qualitative Review of Unconventional Lasers based on Photonic Crystals, LAJPE, LAT. AMERICA 8 (2014) 4307-1–4307-7.

Heller W., Remarks on refractive index mixture rules. J. Phys. Chem. 69 (1965) 3–1129.

Holmström P, Thylén L, & Bratkovsky A. Dielectric function of quantum dots in the strong confinement regime, J Appl Phys., 107 (2010) 064307.

Luan Nannan, Wang Ran, Lv Wenhua, Lu Ying, & Yao Jianquan, Surface Plasmon Resonance Temperature Sensor Based on Photonic Crystal Fibers Randomly Filled with Silver Nanowires, Sensors, 14(2014) 16035-16045.

Peng Y., Hou J., Huang Z., & Lu Q., Temperature sensor based on surface plasmon resonance within selectively coated photonic crystal fiber. Appl. Opt. 51 (2012) 1–6367.

Quach James Q., Su Chun-Hsu, Martin Andrew M., Greentree Andrew D. & Hollenberg Lloyd C.L., Reconfigurable quantum metamaterials, Phys Rev A,80 (2009) 063838.

Sturges T.J., Woollacott C., Weick G., & Mariani E. Dirac plasmons in bipartite lattices of metallic nanoparticles, 2D Mater., 2 (2015) 014008.

Wang R., Wang, Y., Miao, Y. Lu, Y., Luan, N., Hao, C., Duan L., Yuan, C., & Yao J., Thermo-optic characteristics of micro-structured optical fiber infiltrated with mixture liquids, JOSK, 17 (2013) 231–236.

Wang R., Yao J., Miao Y., Lu Y., Xu D., Luan N., Musideke M., Duan L., & Hao C., A reflective photonic crystal fiber temperature sensor probe based on infiltration with liquid mixtures. Sensors 13 (2013) 7916–7925.

Weick G, & Mariani E., Tunable plasmon polaritons in arrays of interacting metallic nanoparticles, Eur Phys J B., 88 (2015) 7.

Weick G., Woollacott C., Barnes W.L., Hess O., & Mariani E., Dirac-like Plasmons in Honeycomb. Lattices of Metallic Nanoparticles. Phys Rev Lett. 110 (2013) 06801.

Advances in Optoelectronic Technology and Industry Development – Jose & Ferreira (eds)
© 2020 Taylor & Francis Group, London, ISBN 978-0-367-24634-1

Ultrafast quantum random number generation based on quantum phase fluctuation unlimited by coherence time

W. Liu & W. Hong*

Wuhan National Laboratory for Optoelectronics, School of Optical and Electronic Information, Huazhong University of Science and Technology, Wuhan, China

ABSTRACT: An improvement for Quantum Random Numbers Generator (QRNG) scheme based on measuring laser phase noise is proposed, where several selectable branches is used instead of two fixed branches as in a traditional Mach-Zehnder Interferometer (MZI). The operation principle behind the improved scheme is discussed and the reason why the sampling rate of the improved scheme can break through that of the traditional MZI scheme without losing system entropy is analyzed. The operation time sequence of the optical switches and the system requirement of the improved scheme are also given. The results show that the limitation of laser coherence time is shifted to the requirement of the delay time in each branches and the sampling rate of the improved scheme can be multiple times of the traditional scheme.

Keywords: quantum random numbers generator, coherence time, phase fluctuation, sampling rate

1 INTRODUCTION

Random numbers play an essential role in many fields, such as cryptography, scientific simulations, lotteries and fundamental physics tests (Ma X F et al. 2016). True Random Numbers (TRNs) can be generated from fundamental optical quantum processes, such as photon counting (Stipcevic M et al. 2007), branching path or time of arrival (Yan Q et al. 2014, Nie Y Q et al. 2014), vacuum fluctuations (Symul T et al. 2011, Gabriel C et al. 2011), amplified spontaneous emission (ASE) (Williams C R S et al. 2010, Li X et al. 2011, Li L et al. 2014) and laser phase noise (Qi B et al. 2010, Xu F et al. 2011, Nie Y Q et al. 2015, Yang J et al. 2016, Liu J et al. 2017, Raffaelli F et al. 2018, Kadhim F A et al. 2018), etc. To date, raw rate of quantum random numbers generation (QRNG) up to Gbps has been reported, only with schemes based on vacuum fluctuations, ASE and laser phase noise.

QRNG schemes based on laser phase noise arouse much research interest due to ultrafast generation speed, commercially available constituent devices and integration potential. Laser phase noise is a non-predictable quantum-mechanical process and can't be measured directly. Mach-Zehnder Interferometer (MZI) is often used to convert phase fluctuation into intensity fluctuation before measurement. Raw data rate of this scheme is limited by the coherence time of laser (Qi B et al. 2010).

In this paper, we report a new scheme for QRNG based on quantum phase fluctuation, which is unlimited by laser coherence time. Section II of this paper describes the structure and principle of our scheme. Section III analyzes and discusses the working flow and system requirement. Section IV is conclusion.

2 STRUCTURE AND PRINCIPLE

According to the literature (Qi B et al. 2010, Zhou H et al. 2015), there are some constraints for the theoretical maximum sampling speed of QRNG schemes based on laser phase noise and subsequent phase-to-amplitude conversion utilizing MZI. Firstly, in order to ensure the independence of the two interfering light fields and thus obtain maximum entropy of the QRNG system, the time delay between the two interfering light fields in the MZI should be greater than the coherence time of laser. That is:

$$\Delta t \geq \tau_c \tag{1}$$

where Δt is the time difference between the two arms of the MZI and τ_c is the coherence time of the laser. Other constraints include (Qi B et al. 2010):

$$T_s > \Delta t + T_r \tag{2}$$

$$\tau_c > T_r \tag{3}$$

T_s is the reciprocal of the sampling rate and T_r is the response time of photo-detection system. These constraints can be combined and rewritten as

$$T_s > \Delta t > \tau_c > T_r \tag{4}$$

As T_r of commercially available photodetectors is larger than coherence time of ordinary lasers, the general constraint (4) indicates that the maximum sampling speed is limited by the coherence time of the laser and the theoretical maximum is obtained when $T_s = \Delta t = \tau_c$.

We first consider the four cases in which T_s, Δt and τ_c satisfy different relationships, as shown in Figure 1 (a–d).

According to the theory of laser phase noise, the sampling result is determined by phase noise from the spontaneous photons emitted in the time window of specified by Δt (Zhou H et al. 2015). Figure 1a indicates that sampling results S_1 and S_2 are totally independent, as the two sampling time corresponds to time intervals without overlapping. The sample rate

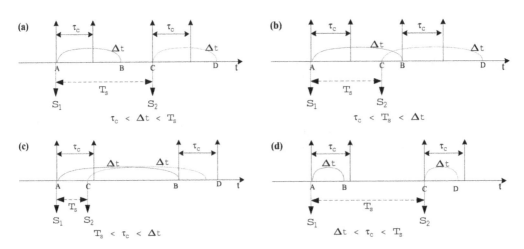

Figure 1. The four cases in which T_s, Δt and τ_c satisfy different relationships: (a) $\tau_c < \Delta t < T_s$, (b) $\tau_c < T_s < \Delta t$, (c) $T_s < \tau_c < \Delta t$ and (d) $\Delta t < \tau_c < T_s$. In these sub-figures, sampling result S_1 is determined by phase noise from the spontaneous photons emitted in the time window of (A, B), sampling result S_2 is determined by phase noise from the spontaneous photons emitted in the time window of (C, D).

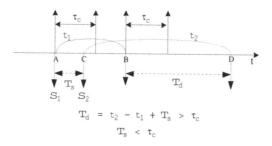

$$T_d = t_2 - t_1 + T_s > \tau_c$$

$$T_s < \tau_c$$

Figure 2. Principle of the improved scheme.

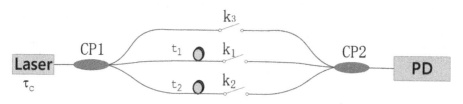

Figure 3. The structure of improved scheme. CP1: optical coupler (1:3), CP2: optical coupler (3:1); τ_c: coherence time of laser; t_1, t_2: delay time of optical delay line in middle and lower branch; k1~k3: optical switches; PD: photodetector.

$1/T_s$ is limited by τ_c in these case. Figure 1b indicates that sampling results S_1 and S_2 are correlated, but this correlation can be eliminated by "XOR" operations. In this case, the sampling results corresponds to the time window specified by T_s instead of $\varDelta t$ (Zhou H et al. 2015). As $T_s > \tau_c$, the entropy of the system can be still guaranteed. Figure 1c is different to 1b in that the time window overlap is quite large, and as $T_s < \tau_c$, the entropy of the system cannot be guaranteed. Figure 1d indicates that although sampling results S_1 and S_2 are independent, the entropy of the system can't be guaranteed as $\varDelta t < \tau_c$.

To break through the limitation of laser coherence time, Figures 1b and 1c are considered comprehensively. The idea of our improve scheme is shown in Figure 2. The difference between Figure 2 and Figure 1c is that the time difference between B and D (T_d) is larger than τ_c, while that between A and C is smaller than τ_c. In this case, the correlation between S_1 and S_2 can be eliminated by "XOR" operations and the entropy of the system can be guaranteed as $T_d > \tau_c$ (Zhou H et al. 2015). Noting that time difference between A and C is just the sampling time T_s, which can be smaller than τ_c. This means that the limitation of generation rate by laser coherent time is overcome.

The structure of the improved scheme is shown in Figure 3. The continuous wave from the laser are divided equally into three branches by the coupler and three optical switches are placed in each branch respectively. Turn-on or off of a branch is controlled by the corresponding optical switch. At any time, only two of the switches are at turn-on state and the other is at turn-off state. Thus the two branches involved in interference can be chosen. In this scheme, the maximum sampling rate will be limited by switching speed of the optical switch instead of the laser coherence time.

3 ANALYSIS AND DISCUSSION

The timing sequence of controlling each optical switch in a whole operation cycle and the corresponding sampling results is as shown in Figure 4. The operation order of each optical switch in one cycle is as follows: k1 and k3 are on, k2 is off, the sampling result after

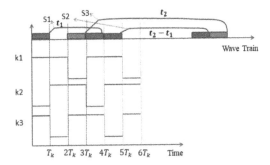

Figure 4. The timing sequence of controlling each optical switch in a whole operation cycle and the corresponding sampling results. T_k: switching interval of optical switch; t_1, t_2: delay time in middle and lower branches of MZI. In the sequence chart of each optical switch, high level (red line) indicates that the optical switch is at turn-on state, while the low level (blue line) indicates turn-off state. On the wave train, two regions in the same color represent the time window for the corresponding to one combination of switching state in the time sequence.

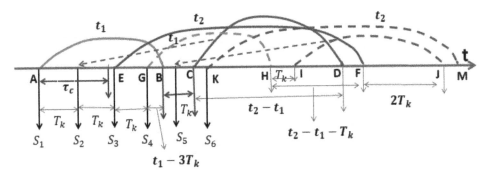

Figure 5. The six sampling points generated in two adjacent switching cycles. T_k: switching interval of optical switch; τ_c: coherence time; t_1,t_2: delay time in middle and lower branches of MZI ($t_1 < t_2$); $S_1 \sim S_6$: sampling results corresponding to time windows AB, CD, EF, GH, IJ, KL respectively.

interference is S_1; then, k1 and k2 are on, k3 is off, the sampling result after interference is S_2; at last, k2 and k3 are on, k1 is off, the sampling result after interference is S_3.

When six adjacent sampling results are generated by using the switching time sequence shown in Figure 4, the sampling results and the time window contributing to the corresponding results are shown in Figure 5.

In Figure 5, the green lines just represent the operation of an ordinary two-arm MZI scheme. In order to guarantee the system entropy while the sample results (after XOR operation) are independent of each other, the time windows contributing respectively to the two adjacent sampling results should ensure that the separation of one end of these windows is greater than τ_c, as shown in Figure 2. According to this condition, the time parameters in Figure 5 should meet the following inequalities:

$$AG: 3T_k > \tau_c \tag{5}$$

$$EC: t_1 - T_k > \tau_c \tag{6}$$

$$CD: t_2 - t_1 > \tau_c \tag{7}$$

$$HF: t_2 - t_1 - T_k > \tau_c \tag{8}$$

$$\text{DH}: t_2 - t_1 - 2T_k > \tau_c \tag{9}$$

$$\text{IF}: t_2 - t_1 - 2T_k > \tau_c \tag{10}$$

$$\text{DM}: 4T_k > \tau_c \tag{11}$$

where AG represents the time window between A and G, etc. By merging inequalities (5)–(11), the constraints on $t_{1,2}$ and T_k can be expressed by:

$$T_k > \tau_c/3 \tag{12}$$

$$t_1 > 4\tau_c/3 \tag{13}$$

$$t_2 > 3\tau_c \tag{14}$$

It should be pointed out that the inequality (5) is strictly required. This condition ensures that S_1 and S_4, or the sampling results in two adjacent cycles, are independent (after XOR operation) and system entropy can be guaranteed. In the case of Figure 3, there are three permutations for selecting two of the three branches to interfere. The maximum sampling rate will be $1/T_s = 1/T_k = 3/\tau_c$ when $t_{1,2}$ meets the inequalities (13) and (14), which is threefold of the theoretical maximum speed of $1/\tau_c$. It should be pointed out that the different on-off timing sequence of the optical switches may results in different requirements for delay time of each branches, which will cause a little difference in the form of inequalities (12)–(14).

Theoretically, even larger sampling rate can be achieved by increasing the number of optical path branches and properly setting the on-off time sequence of the optical switches involved in each branch. If there are n selectable branches in the MZI, the number of permutations of choosing two switches for on state is $n(n-1)/2$. The left-hand side of inequality (5) will be replaced by $n(n-1)T_k/2$, and inequality (5) should be replaced by inequality (15) in this case:

$$\frac{n(n-1)}{2} T_k \geq \tau_c \tag{15}$$

and the maximum sampling rate $n(n-1)/2$-folds of the theoretical value can be expected:

$$\left(\frac{1}{T_s}\right)_{Max} = \frac{1}{T_k} = \frac{n(n-1)}{2\tau_c} \tag{16}$$

From the above analysis, it can be found that the improved scheme does not conflict with existing theories. The raw data generated by the new scheme is theoretically equivalent to that generated by the traditional Mach-Zehnder interferometers. The system constraint equivalent to (4) can be expressed simply by:

$$T_s \geq T_k \geq T_r \tag{17}$$

where the limitation of coherent time in the traditional scheme has been removed. In fact, the limitation of laser coherence time is shifted to the requirement of the delay time in each branches, as given by inequalities (13) and (14) in the case of Figure 3.

4 CONCLUSION

In conclusion, we propose an improvement for QRNG scheme based on measuring laser phase noise, where several selectable branches is used instead of two fixed branches as in a traditional MZI. The improved scheme can break through the limitation of laser coherence time while guarantee the system entropy. The sampling rate can be increased to multiple times of the theoretical value by increasing selectable branches involved in the MZI, while the limitation of laser coherence time is shifted to the requirement of the delay time in each branches. The scheme doesn't conflict with existing theories and promising for practical applications

REFERENCES

Ma, X.F., Yuan, X., Cao, Z., Qi, B. & Zhang, Z. (2016). Quantum random number generation. NPJ Quantum Information, 2(1), 16021.

Stipcevic M. & Rogina B.M. (2007). Quantum random number generator based on photonic emission in semiconductors [J]. Review of Scientific Instruments, 78(4): 45104–45100.

Yan, Q., Zhao, B., Liao, Q. & Zhou, N. (2014). Multi-bit quantum random number generation by measuring positions of arrival photons. Review of Scientific Instruments, 85(10), 103116.

Nie, Y.Q., Zhang, H.F., Zhang, Z., Wang, J., Ma, X. & Zhang, J, et al. (2014). Practical and fast quantum random number generation based on photon arrival time relative to external reference. Applied Physics Letters, 104(5), 051110.

Symul, T., Assad, S.M. & Lam, P.K. (2011). Real time demonstration of high bitrate quantum random number generation with coherent laser light. Applied Physics Letters, 98(23), 145.

Gabriel, C., Wittmann, C., Sych, D, Dong, R., Mauerer, W. & Andersen, U.L., et al. (2010). A generator for unique quantum random numbers based on vacuum states. Nature Photonics, 4(10), 711–715.

Williams, C.R.S., Salevan, J.C., Li, X, Roy, R. & Murphy, T.E. (2010). Fast physical random number generator using amplified spontaneous emission. Optics Express, 18(23), 23584.

Li, X., Cohen, A.B., Murphy, T.E. & Roy, R. (2011). Scalable parallel physical random number generator based on a superluminescent LED. Optics Letters, 36(6), 1020–1022.

Li, L., Wang, A., Li, P., Xu, H., Wang, L. & Wang, Y. (2014). Random bit generator using delayed self-difference of filtered amplified spontaneous emission. IEEE Photonics Journal, 6(1), 1–9.

Qi, B., Chi, Y. M., Lo, H.K. & Qian, L. (2010). High-speed quantum random number generation by measuring phase noise of a single-mode laser. Optics Letters, 35(3), 312–314.

Xu, F., Qi, B., Ma, X., Xu, H., Zheng, H. & Lo, H.K. (2011). An ultrafast quantum random number generator based on quantum phase fluctuations. Issue, 11(11), 12366–12377.

Nie, Y.Q., Huang, L., Liu, Y., Payne, F., Zhang, J. & Pan, J.W. (2015). The generation of 68 gbps quantum random number by measuring laser phase fluctuations. Review of Scientific Instruments, 86(6), 063105.

Yang, J., Liu, J., Su, Q., Li, Z., Fan, F. & Xu, B., et al. (2016). 5.4 gbps real time quantum random number generator with simple implementation. Optics Express, 24(24), 27475–27481.

Liu, J., Yang, J., Li, Z., Su, Q., Huang, W. & Xu, B., et al. (2017). 117 gbits/s quantum random number generation with simple structure. IEEE Photonics Technology Letters, PP(99), 1–1.

Raffaelli, F., Ferranti, G., Mahler, D.H., Sibson, P., Kennard, J.E. & Santamato, A., et al. (2018). A homodyne detector integrated onto a photonic chip for measuring quantum states and generating random numbers. Quantum Science & Technology, 3(2).

Kadhim, F.A. and H.I. Mhaibes, (2018). Quantum Random Bits Generator based on Phase Noise of Laser, J. Eng. Appl. Science. 13, 629.

Zhou, H., Yuan, X. & Ma, X. (2015). Randomness generation based on spontaneous emissions of lasers. Physical Review A, 91(6).

Advances in Optoelectronic Technology and Industry Development – Jose & Ferreira (eds)
© 2020 Taylor & Francis Group, London, ISBN 978-0-367-24634-1

2D light confinement in a MOSFET structure based on near-zero permittivity

S.Y. Sun & W. Hong
Wuhan National Laboratory for Optoelectronics, Huazhong University of Science and Technology, Wuhan, Hubei, China

ABSTRACT: A Metal–Oxide–Semiconductor Field-Effect Transistor (MOSFET) in the inversion mode is investigated as both an electronic device and a photonic device. The permittivity distribution in the semiconductor region is calculated from the electron density distribution for different gate and source–drain voltages. It is found that an inversion layer of electrons formed under the gate electrode can be a near-zero permittivity (Epsilon-Near-Zero, ENZ) region due to the graded distribution of permittivity. By conducting mode analysis, it is further found that 2D confinement of an electromagnetic field can be achieved due to the graded distribution of permittivity in both x and y directions in the transverse plane. The advantage of this structure is that the ENZ performance can be tuned over a wide frequency and the field distribution in the transverse plane can be tailored at a specific frequency by changing gate and source–drain voltages.

1 INTRODUCTION

In recent years, Epsilon-Near-Zero (ENZ) material has attracted increasing attention due to its interesting features, such as near-zero refractive index, decoupling of electric field and magnetic field (Engheta, 2013), having infinite wavelength and phase velocity of the oscillating wave (Liberal & Engheta, 2017), and so on. Many potential applications have been proposed, such as electromagnetic energy tunneling through subwavelength channels (Silveirinha & Engheta, 2006), radiation phase pattern tailoring (Alu et al., 2007; Bravo-Abad et al., 2012; Liberal & Engheta, 2016), optical nonlinearities (O'Brien et al., 2014), super-resolution or low-aberration lenses (Yang, L. et al., 2015; Kita et al., 2017; He et al., 2017), and so on. ENZ can be realized in various continuous media including polaritonic materials, metals and doped semiconductors (Caldwell et al., 2015; Kim et al., 2016; Blazek et al., 2015), as well as metamaterials (Brown, 1953; Maas et al., 2013; Moitra et al., 2013; Goncharenko & Pinchuk, 2014). However, both types of ENZ material encounter the problem that their ENZ function is only specified to a certain frequency range once the material has been designed, which results in difficulties in tunability and controllability of the functional structures based on these ENZ materials. It is reported that linearly doped semiconductors can enhance field confinement in a transition layer where a zero-cross of permittivity exists (Goncharenko & Pinchuk, 2014). It is noted that an ENZ region is introduced near this zero-cross. In this paper, we propose to use a Metal–Oxide–Semiconductor Field-Effect Transistor (MOSFET) in the inversion mode to obtain two-dimensional (2D) graded distribution of permittivity in the transverse plane, thus achieving 2D confinement of the electromagnetic field. The advantage of this structure is that the ENZ performance can be tuned over a wide frequency range by changing gate and source–drain voltages.

2 STRUCTURE AND PRINCIPLE

The structure we studied is a typical N-type silicon MOSFET, as shown in Figure 1. The P-type silicon body is 0.9 μm wide and 0.7 μm thick, with doping concentration of 10^{17} cm^{-3}.

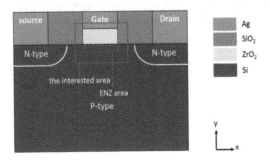

Figure 1. The MOSFET structure.

The oxide layer is 5nm-thick ZrO_2. Gate, source and drain electrodes are made of Ag with SiO_2 as separating material in between. The width of the gate electrode is 0.3 μm, while that of the source the drain electrodes is 0.2 μm. The N-type doping areas of source and drain are both 0.25 μm wide and 0.1 μm thick with N-type doping concentration of 10^{19} cm^{-3}. If positive voltage is applied to the gate, holes will be depleted under the gate and an inversion layer of electrons will be finally formed with the increasing gate voltage, thus forming an N-type channel between source and drain with high electron concentration (Neamen, 2003). The minimum gate voltage (V_g) needed to form an inversion layer is defined as the threshold voltage, V_T. As the electron concentration is gradually reduced from the oxide–semiconductor interface, a graded distribution of permittivity can be realized in the vertical (y) direction. When a positive source–drain voltage (V_{sd}) is applied, the electron concentration in the inversion channel gradually decreases from source to drain, while it decreases from drain to source with a negative V_{sd}. As such, the distribution of electron concentration in the horizontal (x) direction can be adjusted. Due to the permittivity model of semiconductors given by the Drude–Lorentz model, it is obvious that adjusting V_g and V_{sd} can effectively change the permittivity distribution of Si in the channel region near the oxide–semiconductor interface:

$$\varepsilon(\omega) = \varepsilon_r(\omega) + i\varepsilon_i(\omega) = \varepsilon_\infty \left(1 - \frac{\omega_p^2}{\omega^2 + i\omega\gamma}\right) \tag{1}$$

where ε_∞ is the relative permittivity of undoped semiconductor, ω is the angular frequency, γ is the free carrier damping coefficient, and ω_p is the plasma frequency:

$$\omega_p = \sqrt{ne^2 / (\varepsilon_0 \varepsilon_\infty m_c^*)} \tag{2}$$

in which n is the electron concentration, e is the elementary charge, ε_0 is the permittivity in free space, and m_c^* is the electron effective mass.

With a graded distribution of electron concentration in the channel region, a zero-cross of real permittivity ($\varepsilon_r = 0$) can be realized within a certain wavelength range with the longer-wavelength side decided by the minimum electron concentration and the shorter-wavelength side decided by the maximum electron concentration. In this wavelength range, the electron channel of an N-type MOSFET turns into an ENZ region, which is schematically denoted in Figure 1 by the dashed red rectangle. Because of the boundary condition of an electromagnetic field, the electric field can be significantly enhanced in this region. Furthermore, by applying different V_g and V_{sd}, the electron concentration distribution in this region can be changed, which results in a change of permittivity distribution and the position of the zero-cross for a determined wavelength. Thus, electromagnetic field control can be realized by electron control.

3 RESULTS AND DISCUSSION

3.1 *2D distributions of electron concentration and permittivity*

The electron concentration in the transverse plane can be obtained by using a COMSOL Multiphysics semiconductor module, as shown in Figure 2. The left bottom corner of the MOSFET structure is the origin of the coordinate system. Figure 2(1) shows the overall distribution of electron concentration in the Si region when V_g = 6 V and V_{sd} = 0 V. Figure 2(2) shows the electron concentration along x at y = 0.7 µm. The position is represented by a white horizontal dashed line in Figure 2(1). Figure 2(3) shows the electron concentration along y at x = 0.45 µm. The position is represented by a white vertical dashed line in Figure 2(1). It is clear that the inverted electron concentration in the channel can be much higher than the doping concentration of source and drain. Near the ZrO_2–Si interface, the distribution of electron concentration in the y direction increases linearly from the Si body to the ZrO_2–Si interface. The results for different $V_g(V_{sd}$ = 0) are similar, but the maximum value of electron concentration differs.

Applying different V_{sd}, the electron concentration distribution along x can be changed as shown in Figure 3, where y is also set at 0.7 µm. For a fixed V_g = 6 V, it is noted that, when increasing V_{sd} from 0 V, the electron concentration decreases from source to drain, which results in a graded electron density distribution in the x direction, and the gradient is more obvious for larger V_{sd}.

Then the permittivity distribution in the Si region can be calculated by using Equation 1, while the permittivity of Ag, ZrO_2 and SiO_2 can be found from the literature (Yang, H.U. et al., 2015; Wood & Nassau, 1982; Malitson, 1965). Because the region of concern is near the oxide–semiconductor interface, as denoted by the solid red rectangle in Figure 1, the permittivity distribution in this specific region is shown in Figure 4. Figure 4 compares the permittivity distribution in this region with different source–drain voltages, while gate voltage is V_g = 6 V and operation frequency is f = 172 THz. As shown in Figure 4(1), where V_{sd} = 0 V, it is found that the permittivity of the channel changes from negative to positive in the y direction. There are sharp corners at the left and right ends of the channel, which results from the sudden change of electron concentration between the doped source/drain and the channel, as Figure 3 shows. In Figure 4(2), where V_{sd} = 6 V, the permittivity distribution gradually reduces from source to drain (x direction), which is consistent with the change of the electron concentration distribution shown by the corresponding curve in Figure 3. Thus, the MOSFET structure can control the permittivity distribution in two dimensions as we predict.

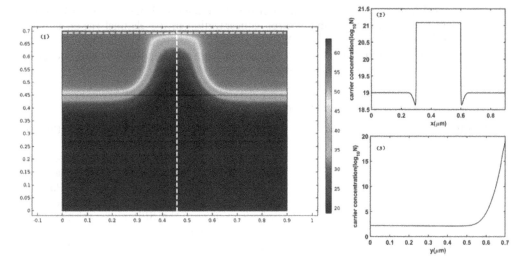

Figure 2. (1) 2D distribution of electron concentration for V_g = 6 V, V_d = 0 V; (2) electron concentration along x, y = 0.7 µm; (3) electron concentration along y, x = 0.45 µm.

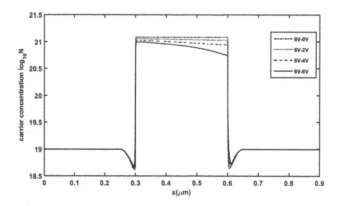

Figure 3. Distributions of carrier concentration in x direction, when $V_g = 6$ V, $V_{sd} = 0$ V, 2 V, 4 V, 6 V.

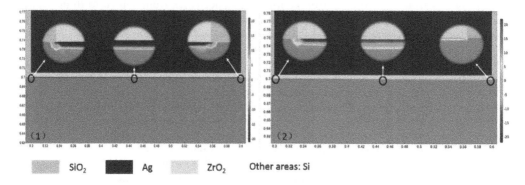

SiO$_2$ Ag   ZrO$_2$ Other areas: Si

Figure 4. (1) Distribution of the permittivity for $V_g = 6$ V and $V_{sd} = 0$ V; (2) distribution of the permittivity for $V_g = 6$ V and $V_{sd} = 6$ V.

After calculating the upper and lower limits of the wavelength range for $\varepsilon_r = 0$ present in the electron channel from electron concentration distribution, the wavelength range for ENZ performance can be determined. As the electron concentration in the channel region is in the range of 3.16^{18} cm^{-3} and 10^{21} cm^{-3} for $V_g = 6$ V and $V_{sd} = 6$ V, the corresponding zero-permittivity wavelength range is from ~0.79 to ~25 μm. This wavelength range may change for different gate and source–drain voltage combinations.

3.2 Electromagnetic field distributions and comparison with SPP mode

In the MOSFET structure, the Ag–ZrO$_2$ interface can support a conventional SPP (Surface Plasmon Polaritons) mode and thus achieve electromagnetic field confinement in the y direction. However, when the permittivity distribution in the N-type channel experiences transition from negative to positive, the region near the zero-cross interface can be analogized as the interface between metal and oxide. As a result, an SPP mode (Goncharenko & Pinchuk, 2014) at this transition interface can be expected. Thus the electromagnetic mode in the whole structure will be different. We compare the dispersion curve of this SPP mode with that of the conventional SPP mode at the Ag–ZrO$_2$ interface, as shown in Figure 5. The result is obtained via a COMSOL Multiphysics RF module. The dispersion curve of the SPP mode with zero-cross interface is almost consistent with the conventional SPP mode, indicating that they are the same mode in essence. Figure 6 compares the distribution of E$_y$ in the y direction under the conditions of $V_g = 6$ V, and $V_g = 0$ V, with $V_d = 0$ V, f = 172 THz. The major concern is

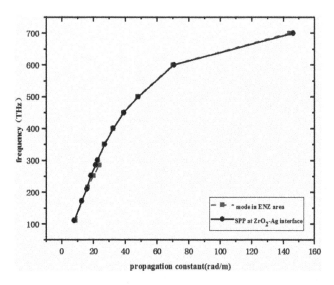

Figure 5.   Dispersion curves of the mode in the ENZ zone when $V_g = 6$ V, $V_d = 0$ V and common SPP mode at metal–oxide interface.

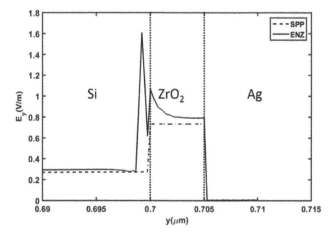

Figure 6.   Field distributions of SPP existing at Ag–ZrO$_2$ interface and mode in the MOSFET structure with the ENZ area for $V_g = 6$ V, $V_d = 0$ V in the y direction, f = 172 THz.

that an ENZ region forms nears the oxide–semiconductor interface when $V_g = 6$ V. For the conventional SPP mode, the field is confined in the oxide layer and decays from the interfaces at both the semiconductor and metal sides. The E_y field with ENZ area that presents near the oxide–semiconductor interface is more confined, because the field peak is larger and field wings are obviously smaller at the semiconductor side and the metal side.

To illustrate the 2D confinement of an electromagnetic field in such a MOSFET structure, the distribution of the E_y field in the transverse plane is shown in Figure 7, where the control voltage is $V_g = 6$ V, $V_{sd} = 0$ V, and the operation frequency is f = 172 THz. Figure 7(1) shows the overall distribution of E_y in the area of interest, which is schematically represented in Figure 1 by the red rectangle. Figure 7(2) shows the distribution of E_y in the y direction as $x = 0.35$ μm. Figure 7(3) shows the distribution of E_y in the x direction as $y = 0.6997$ μm. It is obvious that the E_y field is confined in the ENZ region corresponding to the electron channel. The two peaks and the oscillations of the E_y field at both left and right ends of the ENZ region result from the sharp corners of the permittivity distribution, which is shown in

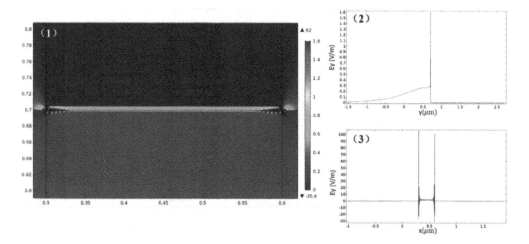

Figure 7. (1) Distribution of electric field (E_y) in the ENZ area, V_g = 6 V, V_{sd} = 0 V, f = 172 THz; (2) electric field (E_y) in the y direction, x = 0.35 μm; (3) electric field (E_y) in the x direction, y = 0.6993 μm.

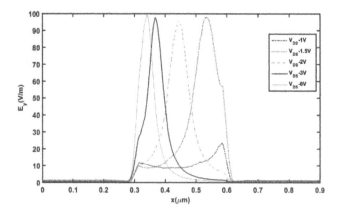

Figure 8. Comparison of the field distributions in the x direction with two conditions, V_{sd} = 1 V, 1.5 V, 2 V, 3 V, 6 V, when V_g = 3 V.

Figure 4(1). The distribution of the E_y field in the x direction can be changed by applying different source–drain voltages, as shown in Figure 8, where V_g = 3 V and V_{sd} = 1 V, 1.5 V, 2 V, 3 V, and 6 V respectively. It is found that the distribution of permittivity in the x direction will change when the distribution of electron concentration varies with V_{sd} increasing, as Figure 4 shows. With larger source–drain voltages, the peak of the electric field (E_y) can be moved from the drain to the source, and the peak is at the very right-hand end of the ENZ region. The results in Figure 8 show that, by applying different source–drain voltages, the field distribution can be tailored in the x direction.

4 CONCLUSION

In summary, we investigate an N-type MOSFET in the inversion mode from both electronic and electromagnetic points of view. It is found that the electron channel near the oxide–semiconductor interface acts as an ENZ region with graded permittivity distribution in both x and y directions. The frequency range for ENZ performance is quite wide due to this graded permittivity distribution and can be adjusted by applying different gate and source–drain voltages. The electromagnetic mode existing in the MOSFET structure with ENZ region presented is found, in essence, to be the same as a conventional SPP mode at the ZrO_2–Ag interface, but one whose field distribution is highly confined in the ENZ region in the y direction. At the same time, the field distribution in the x direction can be adjusted by changing the source–drain voltage, with peak value presenting at the right-hand end of the ENZ region. Hence 2D electromagnetic field confinement can be achieved in a MOSFET structure and the field distribution can be tailored by applying different gate and source–drain voltages. The results suggest the possibility of controlling both electrons and photons with MOSFET structures and a new way to integrate optics with electronics.

REFERENCES

Alu, A., Silveirinha, M., Salandrino, A., et al. (2007). Epsilon-near-zero (ENZ) metamaterials and electromagnetic sources: Tailoring the radiation phase pattern. *Physical Review B Condensed Matter*, 75 (15), 1418–1428.

Blazek, D., Cada, M., Pistora, J. (2015). Surface plasmon polaritons at linearly graded semiconductor interfaces. *Optics Express*, *23*(5), 6264.

Bravo-Abad, J., Joannopoulous, J.D., Soljacic, M. (2012). Enabling single-mode behavior over large areas with photonic Dirac cones. *Proceedings of the National Academy of Sciences*, *109*(25), 9761–9765.

Brown, J. (1953). Artificial dielectrics having refractive indices less than unity. *Proceedings of the IEE - Part IV: Institution Monographs*, 100(5), 51–62.

Caldwell, J.D., Lindsay, L., Giannini, V., et al. (2015). Low-loss, infrared and terahertz nanophotonics using surface phonon polaritons. *Nanophotonics*, *4*(1).

Engheta, N. (2013). Pursuing near-zero response. *Science*, *340*(6130), 286–287.

Goncharenko, A.V. & Pinchuk, A.O. (2014). Broadband epsilon-near-zero composites made of metal nanospheroids. *Optical Materials Express*, *4*(6), 1276.

He, X.T., Huang, Z.Z., Chang, M.L., et al. (2017). Realization of zero-refractive-index lens with ultralow spherical aberration. *ACS Photonics*, *3*(12), 2262–2267.

Kim, J., Dutta, A., Naik, G.V., et al. (2016). Role of epsilon-near-zero substrates in the optical response of plasmonic antennas. *Optica*, *3*(3), 339.

Kita, S., Yang, L., Philip, C.M., et al. (2017). On-chip all-dielectric fabrication-tolerant zero-index metamaterials. *Optics Express*, *25*(7), 8326.

Liberal, I. & Engheta, N. (2016). Nonradiating and radiating modes excited by quantum emitters in open epsilon-near-zero cavities. *Science Advances*, *2*(10), e1600987–e1600987.

Liberal, I. & Engheta, N. (2017). Near-zero refractive index photonics. *Nature Photonics*, *11*, 149–160.

Maas, R., Parsons, J., Engheta, N., et al. (2013). Experimental realization of an epsilon-near-zero metamaterial at visible wavelengths. *Nature Photonics*, *7*(11), 907–912.

Malitson, I.H. (1965). Interspecimen comparison of the refractive index of fused silica. *Journal of the Optical Society of America*, *55*, 1205–1208.

Moitra, P., Yang, Y., Anderson, Z., et al. (2013). Realization of an all-dielectric zero-index optical metamaterial. *Nature Photonics*, *7*(10), 791–795.

Neamen, D. (2003). *Semiconductor physics and devices*. New York, NY: McGraw-Hill.

O'Brien, K., Suchowski, H., Wong, Z.J., et al. (2014). Nonlinear optics in zero index materials. *Lasers & Electro-optics IEEE*, 2160–8989.

Silveirinha, M.G. & Engheta N. (2006). Tunneling of electromagnetic energy through subwavelength channels and bends using -near-zero materials. *Physical Review Letters*, *97*(15), 157403.

Wood, D.L. & Nassau, K. (1982). Refractive index of cubic zirconia stabilized with yttria. *Applied Optics*, *21*(16), 2978–2981.

Yang, H.U., Raschke, M., D'Archangel, J., et al. (2015). Optical dielectric function of silver. *Physical Review B*, *91*(23), 235137.

Yang, L., Shota, K., Philip, M., et al. (2015). On-chip zero-index metamaterials. *Nature Photonics*, *9*(11), 738–742.

Advances in Optoelectronic Technology and Industry Development – Jose & Ferreira (eds)
© 2020 Taylor & Francis Group, London, ISBN 978-0-367-24634-1

A square metal-insulator-metal nanodisk sensor with simultaneous enhanced refractive index sensitivity and narrowed resonance linewidth

Xianchao Liu, Guanhao Cui & Jun Wang

School of Optoelectronic Science and Engineering, University of Electronic Science and Technology of China, Chengdu, China

ABSTRACT: High-performance plasmonic sensors are widely needed in testing refractive index of surrounding environment, identification of gas/solution types and content. Here, a simple square Metal-Insulator-Metal (MIM) nanodisk structure is studied by numerical calculation, in which the metal is Au or Ag. Tunable Fabry-Perot (FP)-like cavity absorption is obtained near the near-infrared for nanodisks composed of Au-insulator-Au or Ag-insulator-Ag, respectively. High absorption remains for a wide range of disk side lengths. The rough sensitivities are high and close to that of the round MIM nanodisks at the same resonance wavelength (Liu & Giessen, 2010). Unexpectedly, the resonance linewidths of the square MIM nanodisk structures are pronounced narrower than that of round MIM nanodisk structures, especially for Ag-insulator-Ag nanodisk structure. Furthermore, when the side length of the insulator disks is scaled down, the refractive index sensitivity (RIS) reaches ~1000 nm/RIU (refractive index unit). The proposed square MIM nanodisk structure is promising for applications in high-performance sensing.

1 INTRODUCTION

Local Surface Plasmon Resonances (LSPRs) (Kristensen & Mortensen, 2017; Willets & Van Duyne, 2007) are optical carrier resonances induced by illumination light on metallic nanostructures. Because LSPRs are easily excited, they have received wide attention in lots of fields, such as broadband absorbers (Yu & Wang, 2019), information storage (Grytsenko & Ksianzou, 2013), molecular sensing (Guerreiro & Sutherland, 2014; Miclăuş & Sutherland, 2016; Ahn & Kim, 2018) and surface-enhanced Raman scattering (Ding & Tian, 2016). The resonance position of a nanostructure depends on its dimension sizes, morphology and surrounding environment. Therefore, fabricated nanostructures can be directly utilized to identify the Refractive Index (RI) of their environment by analyzing peak shifts. Various kinds of nanostructures, such as round metallic disks (Chang & Guo, 2013), triangular metallic disks (Wijaya & Kim, 2017), spherical metallic particles (Navas & Soni, 2014), Metal-Insulator-Metal (MIM) sandwich disks (Liu & Giessen, 2010; Chang & Tung, 2018; Liu & Cui, 2018; Liu & Wang, 2019), have been proposed as LSPR RI sensors. However, RI sensors based on LSPR usually show low rough sensitivity in visible and wide resonance linewidth in the near-infrared, which limit their wider application. Researchers have suggested some strategies and structures for enhancing RIS and narrowing resonance linewidth. Chang and Guo (2013) experimentally obtained Full-Wave-at-Half-Maximum (FWHM) of 10.5 nm by adjusting height and radius of round Au disks. Halas' team showed pronounced narrowing resonance linewidth of Al nanoparticle located on Al film via arousing strong interaction between them (Sobhani & Halas, 2015). Liu and Giessen (2010) reported a high RIS structure, composed of round MIM disks, whose resonance position locates in the near-infrared.

Here, we study a square MIM disks structure, whose LSPR position is adjusted to the near-infrared. Nanodisks, composed of Au-Si$_3$N$_4$-Au or Ag-Si$_3$N$_4$-Ag, are investigated. The side length of a square nanodisk is adjusted and tunable FP-like cavity absorption spectra are obtained for both of them. The FWHM of peaks are much smaller than that of round MIM nanodisks (Liu & Giessen, 2010). In addition, the spectrum sensitivity to environment of the sensors are calculated and shows high RIS. Last but not least, the side length of Si$_3$N$_4$ disk is scaled down and absorption spectrum are calculated. Doubled RIS are achieved. The square MIM nanodisk structure simultaneously shows high RIS and narrow resonance linewidth, which is promising for application in high-performance RI sensing.

2 DESIGN AND NUMERICAL STUDY

The schematic diagram of the suggested structure is shown in Figure 1. The structure is composed of square metal/insulator (MI) disks on surface of a reflective metal mirror (Au-Si$_3$N$_4$-Au orAg-Si$_3$N$_4$-Ag nanodisks structure). Numerical calculation, based on CST microwave studio software was carried out to obtain electromagnetic distribution and absorption spectrum. The unit of MIM disk was put at the center of coordinate space with top and bottom surfaces of structure parallel to xoy plane. The z value of interface between top metal disk and insulator disk is 0 nm. The sides of the squares disks are parallel to the x or y axis. In all calculation, the environment RI was set as 1.33 or 1.33~1.36, which will cover almost all kinds of solutions. The side length of the bottom metal mirror was fixed at 400 nm. The thickness of the top, middle and bottom layers were fixed at 25 nm, 15 nm and 100 nm, respectively. The incident plane wave transmits along the z axis, with electric vector along the x axis. Optical constants were taken from Johnson and Christy (1972) and Palik (1998).

2.1 *Tunable FP-like cavity absorption of square MIM nanodisks*

To compare with round MIM nanodisk structures of Liu and Giessen (2010) and Liu and Wang (2019), side lengths of square MI nanodisks are adjusted to shift the absorption peak to ~1.6 μm. The calculated results are shown in Figure 2. For an Au-Si$_3$N$_4$-Au nanodisk, the side length varies from 120 nm to 200 nm. For an Ag-Si$_3$N$_4$-Ag nanodisk, the side length varies from 130 nm to 210 nm, which is slightly different from that of the Au-Si$_3$N$_4$-Au nanodisk situation. Both square MIM nanodisks structures show tunable absorption spectrum across a wide range. Resonance absorption of FP-like cavity is due to a magnetic resonance located in insulator nanodisks, which arouses the structure response to incident light. From the inset figures, we can see narrow resonance linewidth, especially for Ag-Si$_3$N$_4$-Ag nanodisk, its resonance linewidth can approach 54 nm. The resonance linewidths of round MIM nanodisks reported in Liu and Giessen (2010) and Liu and Wang (2019) are both larger than 200 nm. The narrowed linewidth of resonance may attribute to square disks being better resonance structure than round disk. As

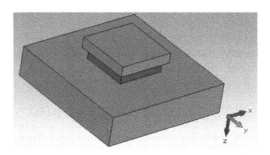

Figure 1. Scheme diagram of the simulation unit.

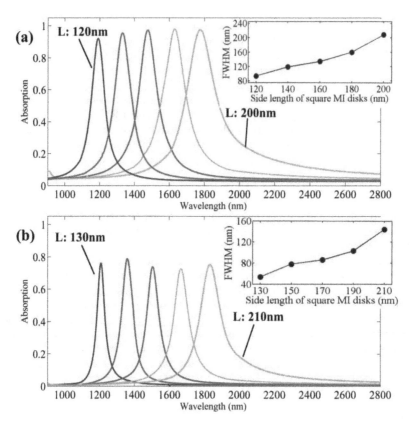

Figure 2. The absorption spectra varying with side length of MI nanodisks: (a) Square Au-Si₃N₄-Au nanodisks structure; (b) Square Ag-Si₃N₄-Ag nanodisks structure. Inset figures are relation between side length of square MI disk and FWHM of absorption spectra.

strong ohmic loss of Au material is near 1.2 µm, the FWHM of absorption peaks of Au-Si₃N₄-Au nanodisks are wider than those of Ag-Si₃N₄-Ag nanodisks.

When side length is 180 nm for Au-Si₃N₄-Au nanodisk and 190 nm for Ag-Si₃N₄-Ag nanodisk, the peak positions of resonances are closer to those of round MIM disks in Liu and Giessen (2010) and Liu and Wang (2019). Therefore, in the following study, we will select the side lengths.

2.2 Spectra response to surrounding environment

Here, spectra response to surrounding environment is studied. The side length of Au-Si3N4-Au nanodisk is 180 nm, and the side length of Ag-Si₃N₄-Ag nanodisk is 190 nm. The resonance peak shift is presented in Figure 3. The environment RI varied from 1.33 to 1.36. The obtained peak shift was 10.8 nm for Au-Si₃N₄-Au nanodisks, and 11.2 nm for Ag-Si₃N₄-Ag nanodisks, corresponding to a RIS of 360 nm/RIU and 373 nm/RIU, respectively. These RIS values are just slightly lower than the RIS result reported in Liu and Giessen (2010), while the Figure Of Merit (FOM) is about twice as large as that of round MIM nanodisks.

3 RIS OF SQUARE MIM NANODISKS BY INCREASING EXPOSED AREA

To enhance RIS of the suggested structures by increasing exposure area of local electromagnetic field, the side length of Si₃N₄ disk was scaled down. Spectra shift with reducing side length of the Si3N4 disk was studied. The shifts of spectra of Au-Si₃N₄-Au nanodisks

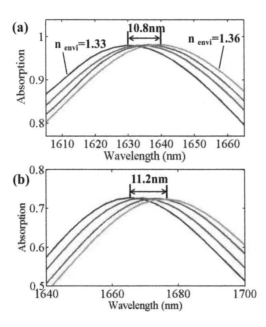

Figure 3. Spectra response to environment: (a) Square Au-Si$_3$N$_4$-Au nanodisks structure; (b) Square Ag-Si$_3$N$_4$-Ag nanodisks structure.

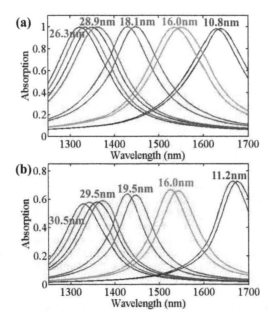

Figure 4. Spectral response to change in the local dielectric environment: (a) Square Au-Si$_3$N$_4$-Au nanodisks structure, the side lengths of Si$_3$N$_4$ are 180 nm, 160 nm, 130 nm, 80 nm and 60 nm, respectively; (b) Square Ag-Si$_3$N$_4$-Ag nanodisks structure, the side lengths of Si$_3$N$_4$ are 190 nm, 160 nm, 130 nm, 80 nm and 60 nm, respectively.

and Ag-Si$_3$N$_4$-Ag nanodisks are shown in Figures 4a and 4b, respectively. Bigger side length of Si$_3$N$_4$ is corresponding to longer resonance wavelength. The same color is selected for a certain side length of Si$_3$N$_4$ nanodisk. For Figure 4a, the side lengths of Si$_3$N$_4$ are 180 nm, 160 nm, 130 nm, 80 nm and 60 nm, respectively. For Figure 4b, the side lengths of Si$_3$N$_4$ are

190 nm, 160 nm, 130 nm, 80 nm and 60 nm, respectively. The peaks shifts of structures with different sizes of insulator nanodisks are marked in the figures. It can be seen that with the side length scale down, RIS increases quickly. When side length is close to 60 nm, RIS of ~1000 nm/RIU is achieved for both Au-Si$_3$N$_4$-Au nanodisks and Ag-Si$_3$N$_4$-Ag nanodisks. These results test that exposure local electromagnetic more to environment is effective for MIM nanostructure to improve sensing performance.

4 CONCLUSION

A square MIM nanodisk structure is suggested for high-perform sensing application. The square MIM nanodisks show well-tunable resonance position and pronounced narrowing resonance linewidths in the calculation results. The resonance linewidth of square Ag-Si$_3$N$_4$-Ag nanodisks is much less than half of those of round MIM nanodisks. By scaling down side length of insulator nanodisk, RIS of the suggested structure increases quickly, which is due to increasing expose area to environment. RIS of ~1000 nm/RIU are obtained by numerical calculation when side length of insulator nanodisk is 60 nm. The suggested structure simultaneously shows narrow resonance linewidth and high RIS and can find opportunity in RI test.

ACKNOWLEDGMENTS

This work was supported by Science Fund for Creative Research Groups of the National Natural Science Foundation of China (No. 61421002), Sichuan Province Science and Technology Innovation Talent Project (No. 2018066), and the China Scholarship Council Fund.

REFERENCES

Ahn, H. & Kim, K. 2018. A localized surface plasmon resonance sensor using double-metal-complex nanostructures and a review of recent approaches. *Sensors* 18(1), 98.

Chang, C.Y. & Tung, Y.C. 2018. Flexible localized surface plasmon resonance sensor with metal-insulator-metal nanodisks on PDMS substrate. *Scientific Reports* 8(1): 11812.

Chang, Y.C. & Guo, T.F. 2013. A large-scale sub-100 nm Au nanodisk array fabricated using nanospherical-lens lithography: A low-cost localized surface plasmon resonance sensor. *Nanotechnology* 24(9), 095302.

Ding, S.Y. & Tian, Z.Q. 2016. Nanostructure-based plasmon-enhanced Raman spectroscopy for surface analysis of materials. *Nature Reviews Materials* 1(6), 16021.

Grytsenko, K. & Ksianzou, V. 2013. Optical recording media based on nanoparticles for superhigh density information storage. *Optical Memory and Neural Networks* 22(3), 127–134.

Guerreiro, J.R.L. & Sutherland, D.S. 2014. Multifunctional biosensor based on localized surface plasmon resonance for monitoring small molecule-protein interaction. *ACS Nano* 8(8), 7958–7967.

Johnson, P.B. & Christy, R.W. 1972. Optical constants of the Nobel metals, *Physical Review B* 6, 4370–4379.

Kristensen, A. & Mortensen, N.A. 2017. Plasmonic colour generation. *Nature Reviews Materials* 2(1), 16088.

Liu, N. & Giessen, H. 2010. Infrared perfect absorber and its application as plasmonic sensor. *Nano Letters* 10(7): 2342–2348.

Liu, X.C. & Cui, G. 2018. Optical properties and sensing performance of Au/SiO2 triangles arrays on reflection Au layer. *Nanoscale Research Letters* 13(1), 335.

Liu, X.C. & Wang, J. 2019. A simple and high-performance platform for refractive index sensing based on plasmonic metal disks on a metal mirror. *IOP Conf. Ser.: Mater. Sci. Eng.* 484 012030.

Miclăuş, T. & Sutherland, D.S. 2016. Dynamic protein coronas revealed as a modulator of silver nanoparticle sulphidation in vitro. *Nature Communications* 7, 11770.

Navas, M.P. & Soni, R.K. 2014. Laser generated Ag and Ag-Au composite nanoparticles for refractive index sensor. *Applied Physics A* 116(3), 879–886.

Palik, E.D. 1998. *Handbook of optical constants of solids.* Academic Press, New York.

Sobhani, A. & Halas, N.J. 2015. Pronounced linewidth narrowing of an aluminum nanoparticle plasmon resonance by interaction with an aluminum metallic film. *Nano Letters* 15(10), 6946–6951.

Wijaya, Y.N. & Kim, M.H. 2017. A systematic study of triangular silver nanoplates: One-pot green synthesis, chemical stability, and sensing application. *Nanoscale* 9(32), 11705–11712.

Willets, K.A. & Van Duyne, R. P. 2007. Localized surface plasmon resonance spectroscopy and sensing. *Annu. Rev. Phys. Chem.* 58, 267–297.

Yu, P. & Wang, Z. 2019. Broadband metamaterial absorbers. *Advanced Optical Materials* 7(3).

Advances in Optoelectronic Technology and Industry Development – Jose & Ferreira (eds)
© 2020 Taylor & Francis Group, London, ISBN 978-0-367-24634-1

Design and nanofabrication of subwavelength grating-based polarizer for visible wavelengths

Zongyao Yang, Bo Feng, Chen Xu, Bingrui Lu & Yifang Chen
Nanolithography and Application Research Group, State Key Lab of ASIC and System, School of Information Science and Engineering, Fudan University, Shanghai, China

Wenhao Li
Chang Chun Institute of Optics, Chinese Academy of Science, Changchun, Jilin, China

ABSTRACT: Light transmittance and extinction ratio are the two most important properties to characterize the performance of grating-based polarizers. Basic research was conducted to study the structural effect of grating line cross section on the polarization characteristics. To maximize both the transmittance and the extinction ratio, the grating materials, the geometrical dimensions and the grating line structure are systematically studied. Then, a shape in the form of a rectangle stacked with a parabolic shape on top was proposed for achieving both high transmittance and extinction ratio. Nanofabrication for subwavelength gratings in Al was carried out. High transmittance over 70% was achieved, but the extinction ratio was still not satisfactory, which was ascribed to the thickness in the fabricated grating not being large enough. However, the measured ratio fits to the simulation result well, indicating that the designed grating parameters can be a good basis for high-quality polarizers based on subwavelength gratings of aluminum.

1 INTRODUCTION

The polarization-related detection of light is currently attracting growing attention due to its broad applications in navigation, water surface detection, and fuzzy communications (Han, 2017; Sarkar, 2011; Guillaumee, 2009), and so on. By enhancing the contrast and collecting the signals from all-round polarization status, polarimetric imaging is able to distinguish the target, the interfering material, roughness, texture, and other physical and chemical properties. Polarizers using subwavelength metallic gratings based on surface plasmons are one of the most important optical components nowadays for polarimetry detection (Gruev, 2010), thanks to the advances in miniaturization of state-of-the-art nanofabrication (Chen, 2015). The incident light component parallel to the grating direction as Transverse Electric (TE) mode is reflected or converted to heat energy consumption, while that perpendicular to the grating direction as Transverse Magnetic (TM) mode is permeable through the surface plasmon polaritons, thus achieving polarization splitting. High-quality polarimetry imaging requires high transmittance as well as high extinction ratios (the intensity ratio of TM/TE). Although polarizers based on subwavelength gratings with high transmittance and high extinction ratios have already been made commercially available, the integration of polarizers into photo-electron detectors for polarimetric detection still remains a big challenge. Recently, we have succeeded in integrating a subwavelength gold grating into InP-based InGaAs sensors of 1.0–1.6 μm (Wang, 2015). However, the extinction ratio is only 18:1, which is far from satisfactory for high-quality imaging. It was found that the parabolic cross section of the grating lines in the polarimetric detectors (Xu, 2018) should be responsible for such a low ratio. This paper reports our systematic study of the structural effect on the polarizer performance as far as the transmittance and the extinction ratio are concerned. In designing the subwavelength gratings, three metals, gold, silver and aluminum, are compared to find the best one for the grating. Then, the duty cycle, the pitch and the height of the gratings are systematically calculated to establish the geometric dimension window for the line width and the height. Finally, the polarizing performance of the gratings with rectangular and parabolic shapes was compared by using the Finite-Difference Time Domain (FDTD) method. Based on a depth of understanding of the structure's effect on the polarization behavior, this work proposed a novel shape for the grating line cross-sectional structure, that is, the combination of a rectangular with a parabolic shape, trying to achieve high performance in both the transmittance and the extinction ratio.

2 DESIGN OF THE SUBWAVELENGTH GRATINGS BY FDTD SIMULATION

Although the final target was to develop integrated polarimetric detectors, basic research was initially conducted for the subwavelength gratings on quartz wafers. The design of subwavelength gratings for visible wavelengths was carried out using the FDTD method, aiming at achieving a transmittance above 70% and an extinction ratio beyond 50:1. In the initial study, a rectangular shape for the line of the cross section of the grating was assumed, as illustrated in Figure 1. The range of grating dimensions, the pitch, the height of the lines, and the duty cycle as design parameters are summarized in Table 1.

Three metals, Au, Ag and Al, were initially compared. The simulation results in Figure 2 indicate that Al should be the best among the three, fulfilling the requirement for a transmittance above 70% and an extinction ratio beyond 50:1 in optical frequencies. Based on this, the optimization of other parameters will be discussed in the following sections.

2.1 *The dimension range for the Al grating*

With Al as the grating material, the line width and the thickness for high transmittance and high extinction ratio are worked out through FDTD simulations. Figure 3 presents the calculated spectra for the transmittance and the extinction ratio with thicknesses from 50 nm to 200 nm in the visible wavelengths. It can be concluded that both the transmittance and the extinction ratio can satisfy the requirement when the thickness is over 150 nm.

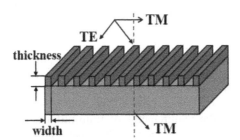

Figure 1. Schematic of the subwavelength grating, with the TM and TE modes.

Table 1. The cross-sectional shapes and the simulation parameters studied in the work; the shape in the last column is a parabolic shape stacked on a rectangular one.

Grating cross section	Al / SiO2	Al / SiO2	Al / SiO2
Material	Al, Ag, Au	Al, Ag, Au	Al
Period	100, 200, 400 nm	100, 200, 400 nm	200 nm
Thickness	50–200 nm	50–200 nm	Case 1 & 2
Duty cycle	0.4–0.6	0.4–0.6	0.4

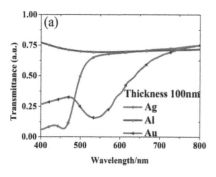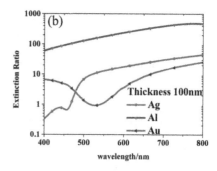

Figure 2. Comparisons of the three metals, Ag, Al and Au, in terms of: (a) Transmittance; (b) Extinction ratios.

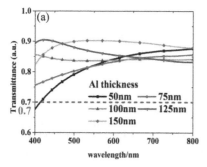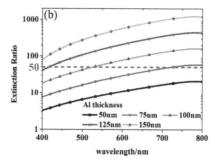

Figure 3. Simulations for the Al grating with a pitch of 200 nm at a duty cycle of 0.4 at various thicknesses from 50 nm up to 150 nm in terms of: (a) Transmittance spectra of TM component; (b) Extinction ratio. The dashed red lines represent the criteria margins for the two measures.

To gain a complete range of parameters for the subwavelength gratings, systematic simulations were carried out to establish the dimension margins, including the pitch, the line width and the thickness. Figure 4 shows the 3D plots in 2D format for both the transmittance and the extinction ratio in a width–thickness plane. The solid lines in the plane define the dimension margins for achieving the requirements of transmittance (>70%) and extinction ratio (>50:1). Such a result is important to guide the design and nanofabrication of grating dimensions for optimal performance of the polarizer.

2.2 The structure's effect on the polarization characteristics

The optimized dimension region shown in Figure 4 was applied in nanofabrication of the subwavelength gratings. However, the actual extinction ratios measured in the fabricated gratings were well below the simulated figures. In inspection of the real gratings, it was found that the real shape of the cross section of the grating lines was parabolic, as shown in Figure 5, rather than rectangular as assumed, suggesting that the polarization characteristics are strongly related to the shape of the grating structure.

To study the structure's effect on the polarization characteristics, both rectangular and parabolic shapes for the grating cross section were compared. Figure 6 presents the transmittance spectra and the extinction ratios for the wavelengths in the visible frequencies. It can be concluded that for the same grating thickness, a parabolic cross section gives rise to a higher transmittance but a lower extinction ratio than the rectangular one does, and vice versa.

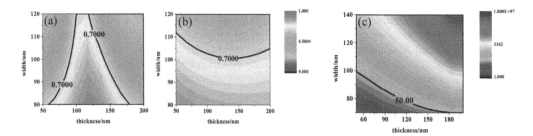

Figure 4. Simulated transmittance of TM component in a grating line width–thickness plane at: (a) λ = 400 nm; (b) λ = 800 nm. (c) Simulated extinction ratio in the width–thickness plane at λ = 400 nm. The cross section of the grating lines is assumed to be rectangular.

Figure 5. Cross-sectional view of the fabricated Al grating, showing the parabolic cross section.

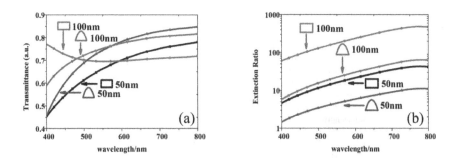

Figure 6. Comparison of rectangular and parabolic grating line cross sections in terms of: (a) Transmittance spectra; (b) Extinction ratios.

2.3 *Structural design for both high transmittance and high extinction ratio*

Having seen the effect of the grating structure on the polarization properties, a novel cross-sectional shape for the grating lines as a parabolic shape stacked on top of a rectangular one, nicknamed as rec-parabolic shape, was proposed in this work, as schematically illustrated in the insets in Table 1. It is anticipated that the parabolic portion of the proposed structure helps to improve the transmittance and the rectangular section enchances the extinction ratio. The heights of both the rectangular and parabolic sections are respectively optimized to maximize both the transmittance and the extinction ratio.

To carry out this process, two different strategies were conducted. One was to fix the rectangle height as 100 nm and vary that of the parabolic section from 10 nm up to 70 nm, to try to meet the polarization performance requirement. The other was to vary the rectangle height while the height of the parabolic section is maintained at 100 nm. Figure 7 presents the calculation results for both the transmittance spectra and the extinction ratio under these two sets of circumstances. It can be seen that as long as the height of the parabolic section or the rectangle is beyond 70 nm and the other part is maintained at 100 nm, the required performance can be readily achieved.

3 NANOFABRICATION

3.1 *Nanofabrication of the rec-parabolic gratings as polarizers*

Nanofabrication of the designed gratings was carried out by state-of-the-art Electron Beam Lithography (EBL) with a JEOL6300 beam writer at an e-beam of 7-nm at 100 keV as the acceleration energy. A bi-layer of 170 nm poly (methyl methacrylate) (PMMA) (MW 100K)/150 nm PMMA (MW 950K) was spin-coated on a clean quartz wafer and then baked in an oven for 60 minutes at 180°C. The exposed PMMA was subsequently developed in a 1:3 mixture of methyl isobutyl ketone (MIBK) and isopropyl alcohol (IPA) at 23°C for 60 s, and finally rinsed in IPA for 30 s. 100 nm aluminum was deposited by thermal evaporation in vacuum with a NANO36 (Kurt J. Lesker Ltd). The process was completed with a wash-off in acetone. Figure 8 presents the fabricated subwavelength grating in Al with the nominal thickness of 100 nm.

3.2 *Optical characterization*

Optical characterization was carried out in a standard setup consisting of an Olympus microscope and Nova fiber spectrometer in the visible band (400–700 nm), and linear polarizers in

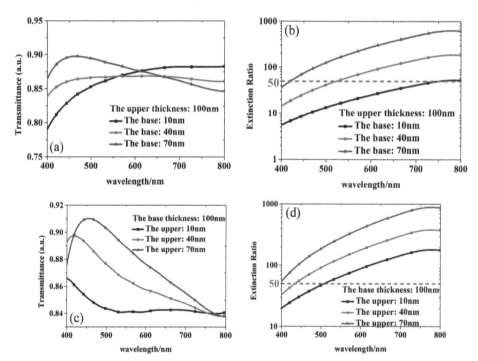

Figure 7.   Thickness optimizations of the rectangular and parabolic parts of the new cross section of the grating lines: (a) Transmittance results when the thickness of the parabolic shape is fixed at 100 nm and that of the rectangular base is varied; (b) Extinction ratio results for same; (c) Transmittance results when the thickness of the rectangular shape is fixed at 100 nm and that of the parabolic section is varied; (d) Extinction results for same.

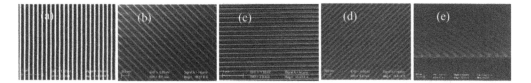

Figure 8.   Nanofabrication results of the subwavelength gratings in Al by EBL and metallization of Al film: (a)–(d) Plan views for four different polarization orientations; (e) Sectional view of the fabricated gratings.

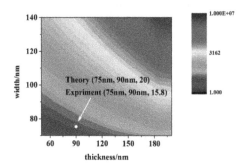

Figure 9.   The measured extinction ratio agrees very well with the simulated result at wavelength 400 nm, according to the white spot (compared with Figure 4c).

serial configuration. A transmittance of around 70% was measured. Unfortunately, the actual extinction ratio of the fabricated polarizer was much lower than expected. Inspection of the Al grating with a ZEISS Sigma HD Scanning Electron Microscope (SEM), shows that the actual height of the grating, illustrated in Figure 8e, was only 75 nm, and the width was 90 nm, which could be responsible for such low ratios. Nevertheless, the measured extinction ratio agrees very well with the simulated result, as shown in Figure 9 by the white spot, indicating that the model established in this work is fundamentally correct. To improve the polarization perform-ance, the thickness of the Al should be significantly increased, to 170 nm as discussed above.

4   CONCLUSIONS

In conclusion, a systematic FDTD sumulation has been carried out in the design of a subwavelength grating for chip-scale polarizers with both high transmittance and high extinction ratio. In the comparison of three metals for the grating material, Al was found the most adequate of the candidates. The geometric margins of the grating structure have been worked out in optimizing the polarizer performance. The study of the structural effect of the grating line shape on the polarization characteristics led us to develop a novel shape for the grating line cross section by stacking a parabolic shape on the top of a rectangular part. Fur-ther simulation study proves that such a structure is able to achieve the desired transmittance as well as the extinction ratio when both of possess sufficient height. A nanofabrication pro-cess by EBL on a double layer of PMMA(MW 100K)/PMMA(MW 950K) was carried out for a 200 nm pitched grating in Al. Over 70% transmittance of the polarizer has been achieved. But the extinction ratio still needs further improvement, by increasing the line thickness. The significant progress achieved in this work points to a trend for further development of grat-ing-based polarizers, which is crucial for the advance of polarimetric imaging techniques.

REFERENCES

Chen, Y. (2015). Nanofabrication by electron beam lithography and its applications: A review. *Microelectronic Engineering*, *135* (C), 57–72.

Gruev, V. (2010). CCD polarization imaging sensor with aluminum nanowire optical filters. *Optics Express*, *18*(18), 19087–19094.

Guillaumee, M. (2009). Polarization sensitive silicon photodiodes using nanostructured metallic grids. *Applied Physics Letters*, *94*(19), 193503.

Han, P. (2017). Active underwater descattering and image recovery. *Applied Optics*, *56*(23), 6631.

Sarkar, M. (2011). Integrated polarization analyzing CMOS image sensor for material classification. *IEEE Transactions on Instrumentation & Measurement*, *11*(8), 1692–1703.

Wang, R. (2015). Subwavelength gold grating as polarizers integrated with InP-based InGaAs sensors. *ACS Applied Materials & Interfaces*, *7*(26), 14471–14476.

Xu, C. (2018). Integrating sub-wavelength polarizers onto InP-InGaAs sensors for polarimetric detection at short infrared wavelength. *Photonics and Nanostructures - Fundamentals and Applications*, in press.

Advances in Optoelectronic Technology and Industry Development – Jose & Ferreira (eds)
© 2020 Taylor & Francis Group, London, ISBN 978-0-367-24634-1

Author Index